Flame Emission and Atomic Absorption Spectrometry

VOLUME 2 — *COMPONENTS AND TECHNIQUES*

Flame Emission and Atomic Absorption Spectrometry

EDITED BY

John A. Dean

DEPARTMENT OF CHEMISTRY
UNIVERSITY OF TENNESSEE
KNOXVILLE, TENNESSEE

AND

Theodore C. Rains

ANALYTICAL CHEMISTRY DIVISION
NATIONAL BUREAU OF STANDARDS
WASHINGTON, D.C.

VOLUME 2   *COMPONENTS AND TECHNIQUES*

MARCEL DEKKER, INC. New York 1971

CHEMISTRY

MARCEL DEKKER, INC.
95 Madison Avenue, New York, New York 10016

LIBRARY OF CONGRESS CATALOG CARD NUMBER: 76-78830
ISBN NO.: 0-8247-1136-X

PRINTED IN THE UNITED STATES OF AMERICA

Preface

The basic principles of flame emission and atomic absorption spectrometry were the topics discussed in Volume 1. Opening the second of this series of volumes is an introductory chapter devoted to the practice and technique of flame methods. It covers a number of miscellaneous topics that are not treated in subsequent chapters and, in the section on "Special Fuels," serves to update material in Chapter 6 of the first volume. Considered next are the basic components of all flame emission and atomic absorption instruments: Light Sources, Burners and Nebulizers, Nonflame Absorption Devices, The Optical Train, Electronics (including detectors), and Instrument Operations. These constitute Chapters 1 through 7 in which both the theoretical and practical aspects are intertwined throughout the discussions. Outlined in Chapter 7 are operational procedures and precautions that are sufficiently general so as to be applicable to available commercial instruments.

Not included in Volume 1 as a distinct separate entity was atomic fluorescence spectrometry. This method, which is still being brought to perfection, forms Chapter 8. Both theoretical principles and technique are discussed.

Instrumentation for flame methods is offered in a variety of forms by many manufacturers. Salient details are gathered together in Chapter 9. In Chapters 10 through 13 are handled the nitty-gritty problems of sample preparation and separation methods, trace analysis and micromethods, evaluation of data, and preparation and storage of stock solutions.

Once again, the editors feel that an outstanding group of authors have agreed to prepare the topical material for this second volume. The editors are grateful to the many instrument manufacturers who so graciously supplied technical information about their equipment and, upon occasion, supplied illustrative drawings and prints.

April, 1971

JOHN A. DEAN
THEODORE C. RAINS

Contributors to Volume 2

Brian W. Bailey, *New York State Department of Health, Division of Laboratories and Research, Albany, New York*

J. A. Brink, *National Physical Research Laboratory, C.S.I.R., Pretoria, South Africa*

L. R. P. Butler, *National Physical Research Laboratory, C.S.I.R., Pretoria, South Africa*

John A. Dean, *Department of Chemistry, University of Tennessee, Knoxville, Tennessee*

D. W. Ellis, *Department of Chemistry, University of New Hampshire, Durham, New Hampshire*

Roland Herrmann, *Department of Medical Physics, University of Giessen, Giessen, West Germany*

H. Massmann, *Institut für Spektrochemie und Angewandte Spektroskopie, Dortmund, West Germany*

Theodore C. Rains, *Analytical Chemistry Division, National Bureau of Standards, Washington, D.C.*

W. G. Schrenk, *Chemistry Department, Kansas State University, Manhattan, Kansas*

Augusta Syty, *Department of Chemistry, Indiana University of Pennsylvania, Indiana, Pennsylvania*

C. Veillon, *Department of Chemistry, University of Houston, Houston, Texas*

Contents

8. Atomic Fluorescence Spectrometry 197

AUGUSTA SYTY

9. Commercial Instruments 235

JOHN A. DEAN

10. Sample Preparation and Separation Methods 263

D. W. ELLIS

11. Trace Analysis and Micromethods 289

ROLAND HERRMANN

Contents of Volume 1

List of Symbols

A	Absorbance, area, Einstein transition probability	I	Intensity
A_t	Total absorption factor	I_A	Intensity absorbed by resonance line
Å	Angstrom unit	I_F	Intensity of atomic fluorescence line
ac	Alternating current		
B	Brightness of source	I_0	Intensity of source
B_x	Factor accounting for decrease in I_0 by absorbing atoms in region x	i	Current, angle of incidence
		k	Absorptivity
		L	Slit length
B_z	Factor accounting for decrease in I_0 by absorbing atoms in region z	M	Metal atom
		M*	Atom in excited state
		MIBK	Methyl isobutyl ketone
B_y	Factor accounting for increase in fluorescence radiation caused by nonresonant absorption	m	Order number
		N	Total number of rulings on grating
		N_0	Total concentration of atoms of interest in a flame
b	Thickness of dielectric spacer, distance between adjacent grooves in diffraction grating	n	Index of refraction
		R	Resistance
C	Capacitance, concentration	R_L	Load resistance
D	Distribution ratio	r	Angle of reflection
dc	Direct current	Rs	Resolution
d	Path length, flame thickness	rpm	Revolutions per minute
E	Potential	rf	Radio frequency
ΔE	Energy difference	S/N	Signal-to-noise ratio
EDTA	ethylenediaminetetraacetate	s	Slit width
emf	Electromotive force	T	Transmittance, effective transmission, temperature
F_T	Total flux transmitting power		
f	Frequency, oscillator strength, focal length	t	Base length of prism
		w	Width, effective aperture width
f/number	Aperture ratio, optical speed		
Δf	Frequency response bandwidth	x	Distance, separation of two emission lines
g	Statistical weight	β	Blaze angle
h	Height	θ	Work function

xiii

λ	Wavelength	$\Delta\nu$	Bandpass, bandwidth (in frequency units)
λ_β	Blaze wavelength		
$\Delta\lambda$	Bandpass, bandwidth, half bandwidth (in wavelength units)	Φ	Fluorescence efficiency
		ϕ	Quantum efficiency
		τ	Rise time (frequency response)
μ	Micro (prefix)	Ω	Solid angle of incident exciting radiation
ν	Frequency		

Flame Emission and Atomic Absorption Spectrometry

VOLUME 2 — *COMPONENTS AND TECHNIQUES*

1 Introduction to the Practice and Technique of Flame Spectrometry

John A. Dean   *and*   *Brian W. Bailey**

DEPARTMENT OF CHEMISTRY
UNIVERSITY OF TENNESSEE
KNOXVILLE, TENNESSEE

NEW YORK STATE DEPARTMENT OF HEALTH
DIVISION OF LABORATORIES AND RESEARCH
ALBANY, NEW YORK

I. Introduction

The basic components of all atomic absorption spectrometers are shown schematically in Fig. 1. Line radiation from a suitable source is passed through an atomic vapor by means of suitable optics to a wavelength selector. The latter isolates the wavelength which can be absorbed by the test material from other spectral lines emitted by the source. A photosensitive detector measures the light flux after passage through the wavelength selector. Usually the output signal requires additional amplification

* Contributed Section II.

1

before presentation to a meter or recorder. Detectors and amplifiers are discussed in Chapter 6. To separate the source radiation from that coming from the excited atoms in the atomic vapor, the spectral source is usually modulated. Atomic fluorescence and atomic absorption require a radiant energy source; one general type is a hollow-cathode lamp. This particular source and several other types of lamps are discussed in Chapter 2.

Flame emission spectrometers dispense with an external source of radiant energy, but in other respects the basic components remain the same

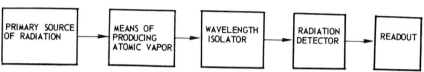

Fig. 1. The basic components of atomic absorption spectrometers. For flame emission work the primary source of radiation is omitted.

as shown in Fig. 1. In the flame emission method the radiant energy emitted by atoms or molecules in the hot flame gases are utilized for their determination.

In all the flame methods, flames are the most popular means for producing an atomic (or molecular) vapor from the test material introduced usually as a liquid aerosol. Common flames, and also some exotic flames, were discussed in Chapter 6 of Volume 1. Liquid fuels and MAPP gas are discussed in Sections II and III of this chapter. The use of a solution as the sample feed provides an ideally "isoformed" sample which can be readily duplicated in standard solutions and in which interferences can be evaluated, eliminated, or compensated for quite readily. The pneumatic means of nebulization (Chapter 3 and Chapter 10 of Volume 1) is likely to remain the basic method of introducing the sample into the burner. And the flame, for all its limitations, is probably unrivaled for both speed and convenience. It undoubtedly will remain the basis of atomic vapor generation for the foreseeable future. Quite generally the type of burner used is intertwined with the choice of fuel and oxidant. In both emission and atomic absorption the flame must be able to dissociate refractory molecular species and convert the analyte into its constituent atoms unless, of course, one is seeking the flame emission of a particular molecular entity. As pointed out in Chapters 11 and 12 of Volume 1, failure to accomplish this step means no atoms for absorption and no atoms available for subsequent thermal excitation and atomic emission or for excitation in atomic fluorescence.

Several different approaches have been tried in order to avoid the use of a flame or the need for dissolution of the sample. Nonflame sources find use in special situations, such as in trace and microanalysis (Chapter 11). Such systems are not as convenient or rapid to operate as the burner and nebulizer systems. Various approaches are mentioned in Chapter 4; most are still in the experimental stages. The use of solid samples is one problem area in which direct vaporization methods are needed. At the same time the use of solid samples restores the problems of inter-elemental and matrix effects, and the need to standardize empirically, using carefully analyzed samples—problems so familiar to spectrographic and X-ray methods.

For some analyses the sample can be nebulized as received or after simple dissolution. Often prior dilution of the test sample is adequate, the degree of dilution being such as to have the signal fall on the linear portion of the calibration curve. In trace analysis a prior concentration step may be necessary. When serious interferences are present, a separation of the test element from other matrix components must precede the flame-spectrometric step. Since a very specific spectrometric step follows, group separations may be completely adequate. Chapter 10 handles the twin problems of sample preparation and separation methods. Evaluation methods are discussed in Chapter 12.

Atomic fluorescence spectrometry is treated in Chapter 8. After the theory is outlined, instrumental parameters, interferences, and operational parameters peculiar to this method are discussed.

Chapter 5 considers the components that comprise an optical train, including auxiliary units. Gratings, prisms, and interference filters are described. When these are incorporated into a monochromator, the optical properties of the latter, such as dispersion, resolution, optical speed, and spectral purity, are discussed. The different requirements of spectrometers and photometers for atomic absorption and atomic emission are brought into focus, as are the characteristics of single-beam, double-beam, and dual-wavelength systems. Salient details of commercial instruments are considered in Chapter 9.

Outlined in Chapter 9 are operational procedures and precautions that are sufficiently general so as to be applicable to available commercial instruments. This chapter should be consulted before any operations are attempted with new equipment. Guidelines for selection of commercial fuels and oxidants follow in this chapter (Section IV). Abnormal signal shapes and their significance are also considered in this chapter (Section V).

II. The Use of Organic Liquids as Fuels in Flame Spectrometry

In the past, workers in the field of flame spectrometry have restricted their choice of fuels to those which could be introduced into the burner systems in the gaseous form. Recent investigations (1–3) have shown that organic liquids may also be used as fuels and that the flames produced from such fuels provide suitable atom reservoirs.

A. FLAME CHARACTERISTICS

1. *Flame Production*

Flames from organic liquid fuels may be produced by aspirating the liquid directly into the mixing chamber of a premix burner system using a

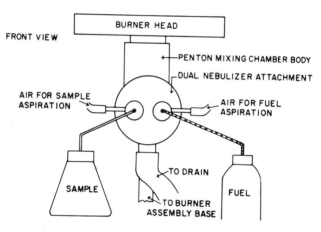

Fig. 2. Dual nebulizer for operation with liquid fuels.

conventional nebulizer. The aerosol thus produced is ignited at the burner face in the normal manner. To introduce a sample into these types of flames a second nebulizer is required; an attachment designed for this purpose is shown schematically in Fig. 2. The composition of the flame may be controlled by either varying the flow rate of the aspirating air or by changing the fuel uptake rate at a constant air flow by using a variable nebulizer. The procedure for obtaining a flame with a minimum of noise (which will be discussed in more detail later) is to set the air flow to a maximum and adjust the uptake rate of the fuel until the required flame conditions are obtained.

2. *Fuels*

At the time of writing relatively few organic liquids have been examined for use as potential fuels. Further, in the investigations that have been carried out the only burner that was used was the Boling 3-slot laminar flow burner. Bearing this in mind, the following organic liquids do not produce stable flames: methanol, ethanol, *n*-butanol, *n*-amyl alcohol, methyl isobutyl ketone, ethyl acetate, amyl acetate, toluene, and xylene. The fuel aerosol either will not burn at the burner face or, after burning erratically for a few seconds, the flame lifts off.

Acetone, *n*-pentane, *n*-hexane, isooctane, and benzene all produce quiet self-supporting flames. With pentane, however, due to its large negative heat of vaporization, the fuel freezes in the nebulizer tip after periods of extended aspiration blocking the nebulizer and causing the flame to extinguish.

3. *Spectra*

The emission spectra of the flames produced with acetone, hexane, and benzene as fuels are as one would expect from air–hydrocarbon flames. These were discussed by Kniseley in Volume 1, Chapter 6 of this treatise. The bands systems associated with the C_2, CH, and OH molecules are all observed. The relative intensities of the C_2 and CH bands are of the same order of magnitude as those emitted by an air–acetylene flame, but the intensity of the hydroxyl band head emission in the region 3046 Å is about two orders of magnitude less for the liquid fuels.

4. *Flame Type*

Stationary flames may be classified under two headings, namely, premixed and diffusion flames. In the case of the former, fuel and oxidant are mixed prior to combustion while, with the latter, the fuel and oxidant are brought together at the site of combustion.

With organic liquid fuel flames the situation is less clear cut. The fuel–oxidant mixture enters the mixing chamber as an aerosol and the type of flame that results is dependent on the extent of vaporization of the aerosol droplets prior to reaching the site of combustion. If the fuel droplets are completely vaporized then the resulting flame is of the premix type.

However, if the air–fuel mixture retains its heterogeneity until it reaches the burner face then the result is a flame of the second type in which each of the aerosol droplets function as a microdiffusion flame. In actual fact the type of flame produced lies somewhere between these two extremes and is dependent on the physical characteristics of the nebulizer used and the flow rates of the air and fuel. For example, with hexane as a fuel and using a Perkin–Elmer variable nebulizer with an air flow to the fuel nebulizer of 1.6 liters/min and a fuel aspiration rate of 2.4 ml/min, the droplet diameter at the nebulizer orifice has a value of 60 μm.* Sufficient vaporization occurs in the mixing chamber so that the flame produced has a well-defined reaction zone and is essentially premixed in nature. If the air flow rate is increased to 2.5 liters/min and the fuel aspiration rate to 4.1 ml/min the droplet size (theoretical) increases to 120 μm. At these flow rates the flame tends to a diffusion flame with the reaction zone becoming less well defined. The increase in the number of droplets in the flame causes an increase in light scattering and hence an increase in the flame noise. Thus, to obtain the quietest possible flame, the air–fuel flow ratio should be kept as high as possible.

5. Flame Temperature

The flames produced with the organic liquids that have been used as fuels are considerably cooler than those produced with gaseous fuel. Temperature measurements made by the sodium line-reversal technique give maximum temperature of 1960°K for the air–benzene flame, 1800°K for air–hexane and air–acetone flame as compared with 1970°K for the air–natural gas flame and 2300°K for the air–acetylene flame (5). The validity of the results for the air–acetone flame is questionable however. The sensitivity obtained in the atomic absorption determination of calcium with this flame is about half that obtained with the air–hexane flame indicating that it is considerably cooler.

6. Flame Velocity

Flame velocity is an important parameter in determining the geometry of the burner. For flames of high burning velocity, such as nitrous oxide–acetylene, a very small burner orifice is required to prevent the flame from flashing back into the mixing chamber. With flames of low burning velocity the problem is to maintain the flame on the burner face at the required air

---

* Calculated from the equation derived by Nukiyama and Tanasawa (4).

and fuel flow rates, and this necessitates use of a burner with a relatively large orifice.

Lift off velocities have been determined for flames with benzene, hexane, and acetone as fuels and at a variety of fuel and air flow rates. The velocities are relatively low, in the range 30–45 cm sec^{-1}, as compared to a lift off velocity of greater than 185 cm sec^{-1} for an air–acetylene flame under the same experimental conditions.

The low velocities make it impossible to produce flames with the standard single-slot burners that are designed for use with air–acetylene. The Boling 3-slot burner may be used for the fuels that have been mentioned. However, recent investigations have shown that it may be preferable to use a single-slot burner with slot dimensions of 100 × 1.5 mm. With such a burner it is possible to obtain flames with methanol and methyl isobutyl ketone in addition to the other fuels previously mentioned.

7. Oxidants

Apart from air the only other oxidant that has been used in conjunction with organic liquid fuels is nitrous oxide with benzene as the fuel. There are two major obstacles encountered in the use of nitrous oxide as an oxidant. The first is the burner configuration. As was mentioned previously, the conventional single-slot burners designed for use with air–acetylene and nitrous oxide–acetylene flames cannot be used with the organic liquid fuels. Further, it is not recommended that a burner with such a large orifice as the Boling burner be used with nitrous oxide as an oxidant. The procedure that has been used consists of aspirating benzene into a nitrous oxide–acetylene flame burning on the standard nitrous oxide burner. The acetylene flow was then gradually reduced and finally shut off.

At this juncture one encounters the second obstacle. The adiabatic expansion of the nitrous oxide as it leaves the cylinder lowers the temperature to such an extent that it causes the fuel to ice up in the nebulizer orifice after a few minutes of operation. This causes a reduction in the fuel flow rate and the flame to flash back. Heating the nitrous oxide should overcome this problem. However, in view of the other difficulties involved it does not seem practicable to consider using nitrous oxide as an oxidant with organic liquid fuels.

B. ANALYTICAL APPLICATIONS

Investigations into the analytical applications of organic liquid fuels have to date been restricted to a very few elements, and, further, no

TABLE 1

ANALYTICAL SENSITIVITIES FOR Cu, Mn, AND Zn IN
VARIOUS AIR–ORGANIC LIQUID FUEL FLAMES WITH
PERKIN–ELMER MODEL 303 AND DUAL NEBULIZER
MIXING CHAMBER ASSEMBLY

Fuel	Fuel aspiration rate, ml/min	Analytical sensitivity[a]		
		Cu(3247 Å)	Mn(2795 Å)	Zn(2139 Å)
Hexane	2.4	0.06	0.06	0.014
Benzene	2.1	0.07	0.07	0.015
Isooctane	2.5	0.08	0.07	0.016
Acetone	4.0	0.07	0.12	0.019
Acetylene	—	0.13	0.09	0.024

[a] Defined as concentration in μg/ml necessary to produce 1 % absorption.

studies have been made on matrix effects. Drawing from previous experiences by other investigators with the relatively cool air–coal gas flame it seems reasonable to say that the liquid fuels will give better sensitivities for some elements because of reduced ionization; however, matrix effects will probably be enhanced.

The sensitivity obtained in the atomic absorption determination of copper, manganese, and zinc in various organic liquid fuel flames are shown

TABLE 2

EXPERIMENTAL CONDITIONS AND ANALYTICAL
SENSITIVITIES OBTAINED FOR Cu, Fe, Ag, Pb, Mn,
AND Zn IN AN AIR–HEXANE FLAME

Element	Wavelength, Å	Slit width, mm	Fuel aspiration rate, ml/min	Sensitivity[a]	
				Air–hexane flame	Air–acetylene flame
Cu	3247	1.0	2.4	0.06	0.13
Fe	2483	0.3	2.4	0.16	0.25
Ag	3281	1.0	2.4	0.11	0.19
Pb	2833	1.0	3.0	0.5	0.5
Mn	2795	1.0	2.4	0.06	0.09
Zn	2139	1.0	2.4	0.014	0.024

[a] Defined as concentration in μg/ml necessary to produce 1 % absorption.

in Table 1. The sensitivities of these metals together with results obtained for iron, silver, and lead in an air–hexane flame are shown in Table 2. For these elements the sensitivities are similar to or better than those obtained with air–acetylene.

The detection limits are in keeping with the sensitivity results indicating that the flames are not appreciably noisier than the air–acetylene flame. Detection limits for copper, lead, and zinc in an air–hexane flame are

TABLE 3

DETECTION LIMITS[a] FOR COPPER, LEAD, AND ZINC IN THE AIR–HEXANE AND AIR–ACETYLENE FLAMES

Metal	Wavelength, Å	Detection limit, μg/ml	
		Air–hexane	Air–acetylene
Copper	3247.5	0.013	0.011
Lead	2833	0.39	0.46
Zinc	2138.6	0.0016	0.0028

[a] Defined as $S/N = 2$.

listed in Table 3. Similar enhancements in sensitivities and detection limits should also be obtained for other elements such as sodium, potassium, and bismuth which have previously been shown to give better analytical detection limits in relatively cool air–coal gas flame than in the air–acetylene flame.

In spite of the fact that relatively little work has been done with organic liquid fuels it would seem that they are adequate substitutes for acetylene for those elements which do not form refractory oxides and for which matrix effects are not serious. Possibly the greatest advantage is the instrumental convenience afforded by the use of such fuels in that it dispenses with the need for cumbersome and potentially dangerous tanks of gas.

III. MAPP® Gas as a Fuel

A search for a less explosive and more easily handled gas that matches the properties of acetylene to a high degree has produced a stabilized mixture of methylacetylene-propadiene-propylene known by the trade name, MAPP gas. Its safety characteristics are similar to those of natural gas and propane. In oxygen its burning velocity is about 240 cm/sec. The strong, characteristic odor of the gas makes accidental leaks readily detectable at the 0.01% level, which is far below the lowest explosive limit (3.4%) in air. Copper or high-copper alloys must not be used for piping because the MAPP gas forms unstable acetylides. MAPP, a liquid in cylinders, will last much longer than acetylene cylinders.

Mansell (6) reported that chemical interferences were severe with the air–MAPP flame and that the hotter oxygen–MAPP flame proved to be too explosive. However the combination of nitrous oxide–MAPP is promising (7). It gives a stiff flame and can be operated with burner heads having slots 5–7 cm in length. Slot width should be kept less than 0.56 cm or flashback may occur.

Experimental results for calcium in the presence of phosphorus have been reported (8). Immediately above the burner, or very low in the nitrous oxide–MAPP flame, analytical sensitivity is relatively low and interference is maximum. At about 0.25 cm above the burner head, sensitivity reaches a maximum and interference is less pronounced. At a height of 1.2 cm, sensitivity has fallen off by approximately 30% from maximum but phosphorus interference has disappeared.

IV. Laboratory Manipulations with Gases and Sample Aspiration

A. SELECTION OF GASES

Generally the least expensive grades of cylinder gases are satisfactory for all analytical work. Exceptions and special precautions are commented upon.

1. Acetylene

Welding grade is usually completely satisfactory although occasionally one may encounter contamination by phosphine. When phosphine is present the flame background will exhibit erratic excursions when scanned from 3500 to 6000 Å, and appears more whitish than normal. A cylinder that is filled during the initial stages of the commercial purification cycle will contain less impurities. If impurities become a problem, an analytical grade of acetylene can be procured.

Acetylene is normally supplied in cylinders as a solution in acetone. A container can lose 50–60 ml of acetone for every cubic meter of acetylene evolved at pressures from 15 to 2 atm. As the tank nears exhaustion the proportion of acetone rises rapidly; therefore, acetylene cylinders should be replaced when the pressure drops below 50 psi (or 3.5 atm).

2. *Air*

Compressed air lines should be fitted with glass wool filters to remove particulate matter and liquid droplets of oil and water. To provide a constant pressure of 20 to 40 psi at the nebulizer jet, a higher pressure, 70 to 140 psi (5 to 10 atm), is maintained in a 200-liter ballast tank; a regulating valve then provides the desired constant pressure.

Compressed air from cylinders occasionally causes difficulty because the breathing grade is sometimes obtained by blending oxygen with pure nitrogen. If the oxygen content is too high, flashback could occur in an air–acetylene laminar flow burner. Compressed-air tanks have to be replaced frequently because of the low oxygen content of air. It is desirable to connect several cylinders to a manifold system.

3. *Hydrogen*

Cylinders of hydrogen require no special comment; however, the rapid rate of consumption of hydrogen gas in burners makes it desirable to connect several cylinders to a manifold system for a constant supply and ready replacement of empty cylinders. Each cylinder is fitted with a 2-stage (2000 psi) regulator and needle valve; the outlet of the latter is connected to the manifold system which contains a rotameter inserted ahead of the burner.

4. *Natural Gas and Coal Gas*

Different sizes of fuel orifices are required for each of these gases, and these differ from those employed for acetylene or hydrogen. Filters should be placed in lines coming from transmission mains to remove particulate matter. Regulation is difficult because gas pressures are low and variable.

5. *Nitrous Oxide*

The medical or anesthetic grade is usually the only grade that is conveniently available in most locations. Tank pressure will remain constant at 800 psi until the cylinder approaches exhaustion; as soon as the pressure falls under 800 psi, watch the pressure carefully and replace the cylinder at 200 psi.

6. *Oxygen*

Commercial cylinder gases are sufficiently pure for flame spectrometry. A tank should be replaced when the cylinder pressure has fallen to 200 psi. The oxygen should be supplied to the burner through a regulator capable of delivering 12 ft³/hr at a pressure of 15–20 psi.

7. *Propane*

Propane is supplied commercially in the form of liquefied petroleum gas (LPG) consisting chiefly of propane itself (70–80%) together with butane and minor amounts of other hydrocarbons. As a cylinder nears exhaustion the gas pressure will drift as the butane content builds up; this drift can easily be avoided by not using up the entire content of the container.

Since propane is heavier than air, a leak will form layers on the floor and persist for some time. Propane should not be used in basement laboratories.

B. REGULATION OF OXIDANT AND FUEL GASES

The constancy and reproducibility of a flame depends on the stability of the gas flows supplying the burner and nebulizer. Regulation is achieved by the use of pressure regulators and rotameters.

Two-stage pressure regulators are preferred; they are made from two single-stage units mounted in series. The first unit is used to reduce the cylinder pressure to the range of 10 to 50 psi, and the second unit is used to bring this pressure down to a value slightly higher than that needed. Final adjustment is made with a needle valve. When acetylene is used, the successive pressures are from 225 psi (full tank) to 15 psi in the first regulator, and from 15 to 1.5 psi for the second regulator. A critical pressure ratio must be maintained between the second stage of the cylinder regulator and the narrow-range pressure regulator included among the controls of the flame spectrometer. If the absolute gas pressure indicated on the gauges of the gas cylinder and the narrow-range regulator are, respectively, P_1 and P_2, the ratio P_1/P_2 for acetylene should be 1.8 or over; for oxygen, 1.9, or over. For example, if a pressure of 11 psi is suggested for the gauge on the spectrometer, the overpressure shown on the gauge of the acetylene cylinder should be at least 20 psi.

Rotameters should be placed in the gas supply lines to measure the gas flow and to monitor the constancy and reproducibility of flow. Rotameters are made from a tapered glass tube whose diameter increases progressively upward in the direction of flow of the gas. A weight, spherical or cone-shape and provided with longitudinal grooves, is located inside the glass tube, and is lifted by the inflowing gas to a height which is balanced by the gas flow. Different gases require individual tapered tubes and spheres. To extend the range of flow measurements, some models contain two weights, the upper made from a plastic and the lower weight a metal sphere. Calibration can be by means of a wet-test meter or, more crudely, by timing the expulsion by the gas of water from a container of known volume (and correcting for the partial pressure of water vapor at ambient temperature).

A knowledge of the individual flow rates of the fuel and oxidant enables the operator to choose various fuel-oxidant mixtures ranging from lean flame mixtures to fuel-rich types of flames. There are instances when the quantity of gas flowing to the burner or nebulizer changes owing to a clogging of some orifice. For the same flow rate, an appreciable change in the gauge pressure to achieve the usual flow rate indicates a partially clogged orifice.

1. *Manifold System for Mixing Air and Nitrous Oxide*

The higher temperature and lower burning velocity of the nitrous oxide–acetylene flame eliminates most of the condensed phase interference encountered with the alkaline earth metals. Its major disadvantage is the loss in sensitivity due to ionization of the easily excited atoms. To overcome this loss in sensitivity, a gas flow system can be constructed as shown in Fig. 3 (*9*). With this system air and nitrous oxide are mixed to provide a temperature range extending from 2300°C for an air–acetylene flame to 2995°C for a nitrous oxide–acetylene flame. A mixture of nitrous oxide and air (80/20) with acetylene sufficient to provide fuel-rich conditions eliminates most of the condensed-phase interferences in the determination of alkaline earths.

C. Disposal of Exhaust Gases

An efficient exhaust system over the burner is desirable for all applications and is essential where toxic or noxious products may be encountered. These latter include the elements Be, Cd, Zn, Hg, Tl, Pb, As, P, Se, Te, and Os, as well as fluorides and all radioactive substances. Acid vapors and nitrogen oxides are lung irritants. A small, squirrel-cage blower, operating with a 5-in. diam duct directly over the top of the flame mantle, provides an inexpensive exhaust system for routine work. For laminar flow burners with longitudinal slots, the portion of the duct directly over the burner is

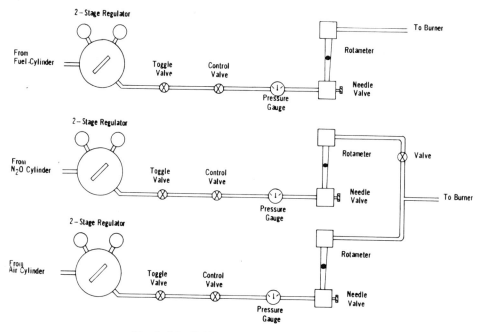

Fig. 3. Manifold system for mixing gases (9).

elongated. Suitable blowers are manufactured by Buffalo Forge Co. as their "baby vent" line; size A has a capacity of 85 ft³/min, and size B about 150 ft³/min. The exhaust gases can be vented into a convenient central hood system or directly outside.

Heat radiating from a flame is itself undesirable in a laboratory. A shield of heat-absorbing glass placed between the operator and burner is a convenient comfort and safety feature should the flame flash back and destroy the laminar flow burner and expansion chamber.

D. SOLUTION FEED RATE

Sample aspiration is another variable that should be properly measured and controlled (see Chapter 13, Vol. 1) both in fundamental investigations and in routine operation. Referred to by various names, such as aspiration rate, sample consumption, solution intake, and feed rate, each is defined by the expression:

$$\text{Feed rate (ml/min)} = \text{Volume consumed (ml)/Time (min)}$$

When the emission intensity or absorption signal is plotted vs feed rate, it usually goes through a maximum. Increasing the feed rate does not have the same effect as increasing the number of atoms by increasing the concentration.

Calculation of feed rate is simple and straightforward. A measured volume of solution is timed for complete uptake. If the solution capillary is held in a fixed position, relative to the sample holder, then allow aspiration to proceed until the solution supply is exhausted. Then pipet a known sample volume into the holder and simultaneously start the timing device; note the elapsed time for consumption of the measured volume. Alternatively, insert the flexible capillary tubing to the bottom of a measured volume of liquid held in a container that is tilted slightly to ensure aspiration of the final portion. Elapsed time is noted.

Instead of measuring the feed rate, the interval between application of the sample to the capillary and the brightening of the flame can be timed; a change in this time interval points to partial obstruction of the nebulizer. Naturally one must aspirate an element whose emission features distinctly color the flame.

V. Signals and Their Shapes

A. CALCULATION OF SIGNAL-TO-NOISE RATIO

Comparison of instrumental performance under different operating conditions involves more than just comparison of signal magnitudes obtained for a given analyte concentration. Noise level should also be considered. Both factors receive their proper emphasis if results are expressed in terms of signal-to-noise ratio as well as signal size (sensitivity).

To calculate the signal-to-noise ratio, the signal is recorded for several instrumental response periods where each response period equals 4 time constants (see Chapter 5, Section IV.A.2). Reject the large, infrequent maxima and minima, and draw a line across the top and bottom of the noise. Measure the span between the top and bottom, as shown in Fig. 4A. Results obtained in this manner are very close to four times the standard

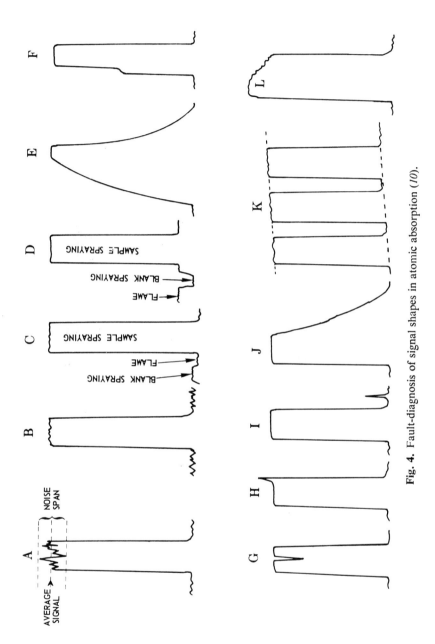

Fig. 4. Fault-diagnosis of signal shapes in atomic absorption (*10*).

deviation (4σ). For small signals (10% or less, expressed in percent absorption) in atomic absorption, the relationship between absorption and absorbance is essentially linear. For signals greater than 10% absorption, conversion to absorbance units is necessary. Then signal-to-noise (S/N) is

S/N = Signal (in absorbance units)/

Peak-to-peak noise (in absorbance units)

B. FAULT DIAGNOSIS OF SIGNAL SHAPES IN ATOMIC ABSORPTION SPECTROMETRY

Knowledge of signal shapes and their causes not only helps the operator evaluate instrument performance according to the appearance of a recorded signal or signals, it also often suggests appropriate corrective action in case of malfunction. Ramirez-Muñoz (10) has published a compilation of recorded signal shapes which emphasize pertinent signal characteristics for laminar flow burners. In many instances similar fault diagnosis would apply to turbulent burners. What follows is abstracted from his series of drawings.

In Fig. 4A there is more noise at the signal than at the blank level. This may arise from an emitting element or may be caused by improper burner alignment in which the source beam traverses the edges of the flame where flame flicker becomes noticable. When the source beam is too high in the flame mantle, the same noise effect may be noted.

In Fig. 4B the noise level is primarily that of the lamp. Absorption reduces the energy seen by the detector and therefore reduces noise for the signal when sample atoms are present in the flame.

When the blank solution absorbs slightly, the situation shown in Fig. 4C results. The opposite case, shown in Fig. 4D, arises when the flame becomes less absorbent when the blank is sprayed. In both instances the instrument would normally be adjusted to read zero absorbance when the blank is aspirated. The absorption readings must be brought on scale in any case so that a difference in percent absorption can be ascertained between sample and blank.

Figure 4E represents a damped signal that can be obtained when a long time constant is used to suppress noise. Should sample aspiration cease before the signal reaches its maximum value, results would be seriously in error.

Formation of gas bubbles within the sample capillary of the nebulizer can give rise to a shoulder in the growth curve (before the bubble is eliminated), as shown in Fig. 4F. A momentary downturn in a signal may also be the result of a brief interruption of aspiration by bubble formation (Fig. 4G).

A short final peak at the termination of the signal plateau (Fig. 4H) may arise because solution remaining in the capillary (after removal from the sample container) feeds more rapidly, producing briefly a higher apparent signal.

A peak at the beginning of aspiration, which is frequently seen in nitrous oxide–acetylene flames, arises from a change in flame transparency when a solution first enters the burner. Too rapid a prefeed rate may also produce a spike preceding the eventual plateau.

Memory effects fall into two categories. After aspiration of a highly absorbing solution, it is normal to get a small peak momentarily when a blank solution or pure solvent is aspirated subsequently. Such a signal is transient in nature and rarely exceeds 5% of the reading of the sample solution. The second category is manifested by a slow return to the zero absorption level, or failure to return completely to the absorbance given originally by the blank. When observed, this indicates the need for thorough cleaning of the nebulizer and burner. Memory effects are shown in Fig. 4I and 4J.

An upward (or downward) drift is disclosed by signals shown in Fig. 4K. Drift is often noticed when using scale expansion. The baseline technique, indicated by the dashed lines, should be used for signal evaluation (see also Chapter 12).

A gradual reduction in signal, noticeable only upon comparison with absorption signals obtained earlier with the same solution, is caused usually by clogging of the sprayer or burner slot, or by change in flame characteristics. Standards should be rerun at frequent intervals to forewarn the operator. A gradual enhancement of the signal indicates a change in flame characteristics; if so, normal behavior can be reestablished by adjustment of the oxidant and fuel gases flow rate.

Several factors may give rise to the type of signal shown in Fig. 4L. After reaching the signal plateau, the signal begins to decrease as spraying continues. The sprayer may be too hot, defective, or off-center. The expansion chamber may be cooling excessively because the aspiration rate is too high or the sampling time too long. In a heated spray chamber perhaps the vapor pressure equilibrium is attained slowly. Decreasing signals may also appear when organic solvents are used.

REFERENCES

1. B. W. Bailey and J. M. Rankin, *Spectry. Letters*, **2**, 159 (1969).
2. B. W. Bailey and J. M. Rankin, *Spectry. Letters*, **2**, 233 (1969).
3. N. S. Poluektov, *Techniques in Flame Photometric Analysis*, Consultants Bureau, New York, 1961, pp. 48–50.
4. S. Nukiyama and Y. Tanasawa, *Trans. Soc. Mech. Eng. Japan*, **4, 5, 6** (1938–1940), Available through the Defence Research Board, Dept. of National Defence, Ottawa, Canada.
5. R. M. Fristrom and A. A. Westenberg, *Flame Structure*, McGraw-Hill, New York. 1965.
6. R. E. Mansell, *At. Abs. Newsletter*, **6**, 6 (1967).
7. A. Hell and S. G. Ricchio, *Flame Notes*, **4**, 37 (1969).
8. A. Hell and S. G. Ricchio, *Flame Notes*, **4**, 41 (1969).
9. T. C. Rains, personal communication.
10. J. Ramirez-Muñoz, *Flame Notes*, **2**, 61 (1967).

2 Light Sources for Atomic Absorption and Atomic Fluorescence Spectrometry

L. R. P. Butler and J. A. Brink

NATIONAL PHYSICAL RESEARCH LABORATORY
C.S.I.R., PRETORIA, SOUTH AFRICA

21

I. Introduction

One of the most important components of atomic absorption and atomic fluorescence instrumentation is the light source. At least part of the remarkable success of atomic absorption as an analytical method must be attributed to the successful manufacture of hollow-cathode lamps and their commercial availability. Since Walsh's (1) first suggestion in 1955 that compact sealed hollow-cathode lamps could be used for atomic absorption, significant advances have followed, and today lamps which radiate the characteristic light of most metallic elements may be purchased at reasonable cost. These lamps are reliable, have extended lifetimes, and due to improvements in electrode design, radiate far more brightly than previous models.

Other types of sources which have been used successfully for atomic absorption are laboratory (arc discharge) vapor lamps, electrodeless discharge tubes, continuous spectral sources, flames, arcs, and sparks. A description of these sources will be given in this chapter.

To understand the processes and mechanisms of these sources, it is necessary that a brief description be given of how electrical conduction through gases takes place. The factors leading to the broadening of spectral lines will also be mentioned, although line broadening generally has been discussed more fully elsewhere (Chapter 3 of Volume 1).

II. Spectral Line Shapes

The shape of the spectral line emitted by the source is of importance both for atomic absorption as well as for atomic fluorescence. The factors which govern the shape of spectral lines are six in number.

Natural broadening is due to the fact that the finite lifetimes of the excited state are not the same for all radiating atoms. Broadening as a result of this is very small and of little importance for atomic analytical spectroscopy.

Doppler broadening is due to the motions of the radiating atoms as a result of thermal activity. For a given atomic line the broadening is proportional to the square root of the temperature. For narrow spectral lines from the source the *temperature* of the radiating plasma should therefore be kept as low as possible. This is done by keeping the lamp current low, or providing cooling either for the electrodes or allowing a fresh cool stream of carrier gas to flush through the discharge.

Lorentz broadening is caused by collisions of excited atoms of *foreign* gas particles, and as not all atoms are affected similarly, broadening and line shift result. Where the gas pressure of the discharge is low, Lorentz broadening is relatively small.

Holtzmark (or Resonance) broadening is also a collisional effect but is due to collisions of radiating atoms with atoms of the *same type*. At low atomic pressure or low concentrations of atoms, these broadening effects are also relatively small (2). Lorentz and Holtzmark broadening together are often called *pressure broadening*. By operating sources used for atomic absorption and fluorescence at reduced pressures, *pressure broadening* effects are minimized, as are shifts in the wavelength peaks of the lines.

Stark broadening is caused by nonuniform electrical fields perturbing the energy levels of emitting atoms. In sources where strong electrical field gradients are present, e.g., high voltage sparks, Stark effect broadening may be appreciable. In the negative glow region of the hollow-cathode lamp, however, electrical potential gradients are small and, consequently, spectral lines originating in this region show little Stark broadening. Stark broadening will cause atoms with low mass numbers to be affected more than those with high mass numbers.

Self-absorption broadening is due to absorption of radiation by non-radiating atoms *in the source*. For sources of extended depth or with significant temperature gradients such as flames, arcs, and Geissler-type discharge lamps, broadening depends on the length of the nonemitting cloud through which the radiation must pass. This type of broadening is evident only in emission sources and can be reduced by shortening the path length and the concentration of vapor through which the light must pass. Absorption broadening is one of the most important factors in lamps used for atomic absorption and fluorescence.

In the hollow-cathode lamps used in atomic absorption, broadening effects that reduce sensitivity most are Doppler and self-absorption broadening. With most modern hollow-cathode lamps, broadening effects have been reduced by good lamp design.

III. Requirements of Sources

Because of the differences in mechanisms of atomic absorption and atomic fluorescence phenomena, there are some differences in the source requirements. Figure 1 shows the techniques of measurement schematically

(a) ATOMIC ABSORPTION

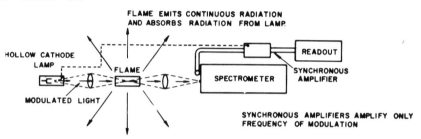

(b) ATOMIC FLUORESCENCE

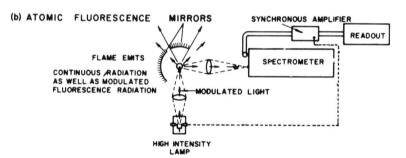

Fig. 1. Schematic diagram showing the measuring techniques for atomic absorption and atomic fluorescence spectrometry.

illustrated. The requirements of sources are here summarized:

1. For atomic absorption

   (a) The source should radiate the light of the element to be determined without interference from other spectral lines originating from impurity elements, electrode materials, or carrier gases (i.e., having wavelengths so close to the resonance lines of the analytical element that they are not resolved in relatively low dispersion spectrometers).

   (b) The resonance spectral lines should be sharp and bright against a very low background.

(c) Radiation intensity must be high for low lamp currents.

(d) The light emission from the sources should be stable and have a constant electrical supply.

(e) Ignition and burning voltages should be low, especially as in some systems the lamp is switched on and off as frequent as 500 times per sec.

(f) The lamps should be small and robust, operate over a considerable period, and, in addition, should have a long shelf-life. Some early lamps were found to deteriorate rapidly when they were not used often, and manufacturers recommend running lamps for a short period at regular intervals to prolong shelf-life.

(g) Physical dimensions and electrode configurations in different lamps should be similar, to facilitate alignment when lamps are interchanged.

(h) The cost of lamps should be reasonable.

2. For atomic fluorescence (see Fig. 1b) the source requirements are similar to those for atomic absorption except that:

(a) The source need not radiate such narrow spectral lines.

(b) The main requirement is a high intensity of the resonance line at the absorption wavelength *peak*. Fluorescence sensitivity is a function not only of atomic concentration in the flame, but also of the intensity of the exciting radiation; the more intense this is, the more sensitive the method (3).

(c) Purity of spectrum is not as important here as it is for atomic absorption because the atoms in the flame will absorb only those wavelengths that cause them to fluoresce, but if scattering of light occurs in the flame, spectral lines which are not resolved by the spectrometer and which are scattered in the flame can lead to erroneous measurement of the fluorescence signals.

IV. Conduction of Electicity through a Gas

As most sources used for atomic absorption and fluorescence spectrometry rely on the conduction of electricity through gases for exciting the spectra, a brief discussion will be given on how this excitation takes place. For those readers more interested in the complexity of gas discharges, several excellent treatises are available (4–6).

An electrical current conducted by a gas between two electrodes is carried by electrons and ions. Electrons liberated at the cathode are accelerated toward the anode by the electrical field. In the process they undergo collisions with other gas atoms causing further ionization and releasing secondary electrons. Ionized atoms, usually being positively charged, move toward the cathode where they produce additional electrons and with their greater mass cause cathodic material to be ejected from the electrode surface. The average distance that a particle can travel in a gas before colliding with another particle (mean free path) depends on the gas pressures and, consequently, the nature of the discharge will depend markedly not only on the gas pressure but also on the properties of the gas carrying the current. The mean free path generally differs considerably for electrons and gas particles, especially at lower gas pressures where an electrical potential exists (nonthermal equilibrium). Because of the collision processes taking place, energy is transferred from those particles being accelerated as a result of the electrical field to noncharged particles. The absorbed energy may then be released in the form of characteristic light, and the intensity of the radiation from a particular atomic species thus depends on the number of atoms present and on the number of collisions taking place. When the collisions are highly energetic, generally because of higher accelerating voltages, spark-like or ionic spectral lines will result; whereas with low energy collisions, atomic and resonance spectral lines will be excited.

A. PRESSURE EFFECTS

The effects of pressure on a dc discharge between two electrodes in a tube are well known. When the pressure is high (above about 20 Torr) discharge is essentially an arc. As the pressure is reduced, the discharge changes into a glow "discharge." At a pressure of about 5 Torr, depending on the carrier gas, the cathode is surrounded by a luminous glow known as the *negative glow*, with a dark space known as the Faraday dark space separating it from the highly luminous positive column. The length of the positive column depends on the distance between anode and cathode, whereas the dimensions of the other regions of the discharge are independent of these dimensions. As the pressure is reduced to approximately 1 Torr, another luminous and dark region is observed. These are termed the cathode glow and Crookes dark space, respectively. Figure 2 shows the regions diagrammatically. At a pressure of about 0.1 Torr the luminous and dark spaces increase in size, and striations may also appear in the positive column depending on the gas type and pressure. The light and dark spaces adjacent to the cathode are attributed to regions where electrons are accelerated (dark spaces). The electrons then achieve velocities so that energy can be interchanged with other particles, collision and excitation leading to luminosity (light spaces). The Faraday dark space is not clearly defined and diffuses into the positive column. A narrow dark band is seen close to the cathode and is known as the Aston dark

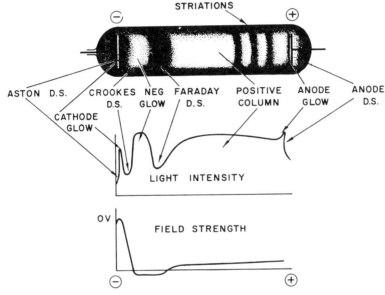

Fig. 2. Diagram of the glow discharge.

space. This dark space is due to the fact that electrons released from the cathode have not yet achieved enough kinetic energy to cause emission of light.

At even lower pressures (less than 0.01 Torr) the cathode glow fades, the luminous positive column disappears, and the discharge would appear to be extinguished except for a faint blue glow on the tube walls.

B. Electrical Effects

The current-voltage graph of a cylindrical discharge tube operating at a constant (low) pressure is shown in Fig. 3. The low current which passes without creating luminosity (A-B) is known as the Townsend discharge. At B the current increases rapidly with a resultant drop in the voltage across the electrodes (C-D). The discharge becomes luminous and shows the glow discharge characteristics mentioned previously. When the current is low the cathode glow does not cover the whole cathode and the discharge is termed the normal glow discharge. It is important to note that, if the diameter of the tube is large in relation to the anode-cathode distance,

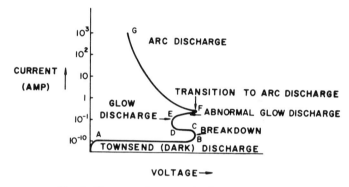

Fig. 3. Current-voltage graph of a glow discharge.

most of the applied voltage appears between the cathode and the boundary between the negative glow and the Crookes dark space (see later). The voltage is constant for current increases under normal glow discharge conditions. A modified version of the glow discharge forms the light emitting region of the hollow cathode. In the negative glow region electrons from the cathode lose energy by collision with gas and metal atoms after having been accelerated through the Crookes dark space. Excitation of these atoms thus takes place here.

As the current increases the whole cathode becomes covered with the glow and the potential again rises (E-F). This is known as the abnormal glow discharge. As the current and voltage rise (E-F), the discharge gives way to the *arc discharge* (F-G), with an uncontrolled rise in current.

The potential difference between the ends of the positive column is relatively small (see Fig. 2) and depends on the diameter of the tube. If the diameter of the tube containing the positive column is reduced, the potential difference rises. Low energy excitation of atoms that have reached this region results in their resonance and arc-like spectra predominating. This characteristic is employed for the selective excitation of resonance lines in high intensity hollow-cathode lamps. It is also used for the excitation of the alkali metal spectra in vapor discharge lamps (see later).

The characteristics of the normal glow discharge are utilized for the excitation of all types of spectra for emission spectroscopy. Certain glow discharge sources are better suited for atomic absorption while others are better suited for atomic-fluorescence spectrometry.

V. Hollow-Cathode Lamps

The hollow-cathode discharge tube has been known since 1916 when it was used as an emission source by Paschen (7). Since then it has been used by many spectroscopists, e.g., Schüler (8), for spectral studies and studies of low pressure discharge phenomena. The properties of narrow spectral lines, high intensity, and the ability to control the discharge critically make it a source preeminently suitable for these purposes.

Although Walsh was not the first to suggest making sealed tubes, it was largely through his instigation that sealed tubes with long lives were made commercially. The tremendous popularity of atomic absorption as an analytical method and the competition among manufacturers have resulted in the high quality lamps that are now available for many elements.

A. MECHANISM OF THE HOLLOW-CATHODE DISCHARGE

The hollow cathode differs from other discharge lamps mainly by virtue of the cylindrical hollow shape of the cathode. It fulfils to a large degree the requirements of atomic absorption, although the intensity of a normal hollow-cathode discharge is generally insufficient to excite fluorescence spectra satisfactorily.

The tube (Fig. 4a) is evacuated and filled with an ultrapure monatomic gas, usually argon or neon. Should a molecular gas or a gas with low ionization potential be present, a considerable portion of the discharge energy would be absorbed by the molecules, resulting in reduced intensity of cathodic metal radiation as well as the presence of molecular band spectra. The hollow cylinder, or cuplike cathode, gives rise to a particular type of low pressure discharge. The positive column, usually predominant in the normal glow discharge, virtually disappears. The inside of the cathode is filled with the negative glow and forms the most important part of the discharge. Between this negative glow and the cathode the negative

(Crookes) dark space is seen (Fig. 4b), and virtually the whole applied voltage falls between this region and the cathode.

As with any glow discharge the current is carried by metal and gas ions and electrons. Positive ions, strongly accelerated through the high electrical gradient of the Crookes dark space, bombard the cathode removing material by sputtering. The metallic particles enter the discharge region, and atoms are excited by collisions with gas ions and electrons which have

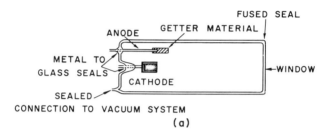

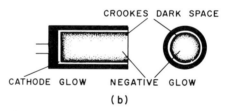

Fig. 4. (a) Diagram of hollow-cathode lamp. (b) Discharge region of the cathode.

been slowed down enough by previous collisions to excite atomic spectra. In the negative glow region these atomic as well as lower energy ionic lines of the cathode material and the carrier gas are strongly visible.

The region of the cathode glow is known to have a very low potential gradient in spite of the high electron and ion densities. Broadening due to Stark effects is thus minimal in this region.

The low gas pressure and relatively low metal vapor density (in comparison to sources operating at atmospheric pressures) result in both types of pressure broadening being small. When low lamp currents are used the temperature of the negative glow plasma is low and Doppler broadening is minimized. However, should the lamp current be increased, both the number of collisions (and thus the temperature) and the rate of sputtering increase, resulting in broader lines due to Doppler broadening and self-absorption broadening. The sputtering rate depends on the cathode material, on the mass of the carrier gas ions, and the distance between anode and cathode (6). Electrons released from the cathode are accelerated through the Crookes dark space to enter the negative glow region. Apart from collisions, fast electrons experience no resistance in this region due to the low electrical field, and so penetrate to the Crookes dark space on the other side. Here they experience an opposite force and their path is reversed. The net result is a cumulative ionization process in the region of the cavity. Gas ions, strongly attracted by the high potential gradient in the Crookes dark space, then bombard the cathode surface.

The hollow-cathode effect is also seen when, instead of a cuplike cathode, parallel plates connected to each other are used as the cathode. Cathodes of this shape have enabled probes to be inserted in the negative glow region to measure the distribution of the charge and current densities. Sturges and Oskam (9) have used such cathode configurations to study the various theories put forward to explain the hollow-cathode effect.

B. The Carrier Gas

The carrier gas, sometimes called the "fill gas," plays a threefold role, namely:

(1) It is largely responsible for carrying the current.
(2) It causes the sputtering action from the cathode.
(3) It is mainly responsible for exciting atoms to radiate characteristic light.

The physical characteristics of the gas play an important part in enabling it to fulfill its purposes most efficiently.

Table 1 gives some of the more important constants of the rare inert

TABLE 1

CONSTANTS OF THE RARE INERT GASES

Gas	Atomic mass, amu	Ionic radius, Å	First ionization potential, eV	Second ionization potential, eV
Helium	4.00	0.93	24.59	54.42
Neon	20.18	1.12	21.56	40.96
Argon	39.95	1.54	15.76	27.63
Krypton	83.80	1.69	14.00	24.37
Xenon	131.30	1.90	12.13	21.21

gases. Both sputtering and excitation are essentially collision processes. Mass and velocity of the ions and the elasticity of the collision will therefore be of importance. From Table 1 it is seen that as the atomic mass increases, so does the ionic radius, but the ionization potential decreases. The ionization potential will largely determine the elasticity of a collision, and thus the transfer of energy from the gas ion to either the cathode walls or to an atom of the cathodic material in the sputtered cloud. As the action of the sputtering process is not clearly understood, the ionic radius of the gas ions may play a role.

Many experiments have been conducted by various workers to determine the optimum conditions of pressure and current for various gases. It would be expected that, because of its high excitation and ionization potential, helium would be a more successful gas for the excitation of spectra. Due to its small ionic radius, however, it is easily adsorbed onto the tube walls by sputtered material and "cleans up" more rapidly. Neon has been found to be one of the most suitable gases for use in hollow-cathode lamps. It is significant that most manufacturers now sell lamps with neon as the carrier gas for most elements, whereas a few years ago argon was generally used. The high ionization potential of neon results in ionic spectra being more intense than is the case with argon. Mitchell (*10*) has shown that the neutral atomic spectra of iron are more intense than the ionic spectra at pressures of neon less than 5 Torr. As the pressure increases the ionic spectra are excited more strongly.

Argon excites more atomic spectra (ratio of iron II/iron I lines is much smaller with argon than with neon). The overall intensity of atomic lines is lower with argon than with neon. In some lamps argon is to be preferred because of the fact that neon spectral lines may interfere with the analytical line of the element. Generally neon has a cleaner spectrum with less interfering lines. With the development of smaller cathodes and their shielding, sputtering has been reduced and the lifetimes of neon-filled lamps extended. It is not uncommon to find modern lamps with lifetimes in excess of 1000 operating hours. The pressure of the gas in the lamp is also important. At lower pressures intensity is generally higher, but sputtering rates and the temperature of the cathode increases because of higher discharge voltage and longer mean free paths. At higher pressures the discharge may become unstable and the intensity is generally lower. The cathode size also plays a role in the optimum gas pressure. When neon is used, gas pressures are of the order of 2–5 Torr, whereas with argon the pressures may vary between 1 and 4 Torr (*11*).

C. LAMP DESIGN AND CONSTRUCTION

Although there are numerous designs of hollow-cathode lamps, only those designs that have been used for atomic absorption and fluorescence spectrometry will be discussed. Earlier commercial sealed lamps made use of a round bulb with two tubular extensions (*12*). One of these held the open-ended cylindrical cathode standing vertically, while the other was provided with the quartz window sealed onto it with pycene wax. These lamps often failed because of gas leakage or contamination. Their design and construction also did not lend themselves to transportation.

The tubular design (see Fig. 4a) has certainly been the most successful for various reasons, viz:

(1) The cathode, having one end closed, gives higher intensity than the double open-ended cathode.
(2) Only a very short support on the closed end of the cathode is needed so that the cathode is rigidly mounted. If the cathode is directly attached to the metal-to-glass sealing wire, this does not induce too much strain on the seal.
(3) Use of a tubular construction simplifies manufacture and reduces costs.
(4)' Mounting, storage, and transport of tubes are much simpler.

There are several types of hollow-cathode lamps available. Because of the commercial competition some design and constructional features are jealously guarded. The descriptions here, therefore, are of a general nature and are not meant to describe any one make, rather describing the designs that have appeared in the literature as well as some used in the authors' laboratory.

Vacuum-tight metal seals are vital. The electrical connections to the electrodes are sealed to the glass. As these electrodes become hot it is essential that the expansion rates of both the metals and the glass to which they adhere are similar. If this is not the case, tiny cracks invisible to the eye may lead to lamp failure through slow leakage of atmospheric gas into the lamp. It is possible to overcome this difficulty by using metal alloys and certain glasses as intermediates between the metal and the tube glass and which match the expansion rates of the metal over a certain temperature range. Figure 5 shows how the coefficients of expansion of glasses and metals change with temperature. It is preferable that the graphs

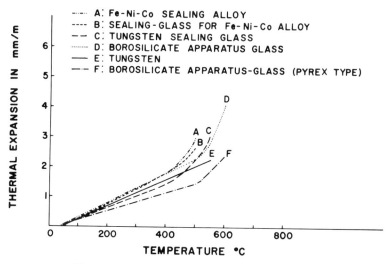

Fig. 5. Expansion coefficients of glasses and metals.

of the glass and the metal used be identical for the temperature to which they are to be exposed.

1. *Electrode Design*

It has been mentioned previously (*13*) that the shape of the anode electrode is not important for stability. The proximity of the anode to the cathode is important, however. Where the electrode material of the anode is zirconium, titanium, tantalum, or some other metal which exhibits "gettering" properties, it is possible to obtain a larger surface area by making the anode circular in shape. In addition, if the getter is to be conditioned under vacuum (i.e., heated to give the required temperature to dry off absorbed gases), it is easier to do this when the shape is circular and an induction coil heater is used. The getter may also be introduced on a separate electrode. It is common practice to encase the anode in a tube to force the discharge to the place where the getter is situated and to maintain stability.

The design of the cathode has received much attention in recent years (*14*). Earlier lamps had cathodes of internal dimensions of 7–12 mm. In recent years cathodes have been reduced in diameter, resulting in much brighter emission intensities for relatively lower lamp currents. Clearly the reduction in cathode diameter must lead to an increase of ion current density per unit volume, and particularly to an increase in electron density, and thus electron temperature within the cathode for the same lamp current. The closer proximity of opposing Crookes dark spaces also leads to higher energy of electrons. Reducing the diameter of the cathode reduces the depths of the Crookes dark space (9). The potential gradients across this region are accordingly increased.

Another means whereby the cathodes are caused to radiate more intensity is to provide a nonconductive shield around the outside of the cathode, as shown in Fig. 6. This shield, made of glass or ceramic material

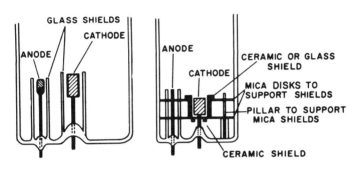

Fig. 6. Diagrams of shielded hollow-cathode lamps.

and being close to the cathode, reduces or prevents a spurious discharge to the outside of the cathode. This more effectively distributes the current to the useful region of the cathode. Consequently, higher light intensities may be obtained from shielded cathodes than would be the case with open cathodes if the same lamp current is passing through the tube. In addition it reduces the sputtering of cathodic material from the outside of the cathode. As the lifetime of a well-made lamp is limited mainly by the decrease in carrier gas pressure or clean up, reducing the sputtering rate effectively extends the lifetimes of lamps. The smaller cathode also enables a smaller image to be focused in the flame, thus utilizing that portion of the flame that is most sensitive. This has resulted in a measurable improvement in sensitivity and precision for certain elements. Smaller cathodes also lead to more light entering the spectrometer and thus lower amplifier gain-settings.

Cathode construction differs considerably for the various metals. Figure 7 shows some typical types of cathodes. Where the metal is easily worked and not particularly expensive, it is usual to manufacture the whole cathode from this metal, as in Fig. 7a. In some instances the metal is very expensive, e.g., platinum, and it is usual to insert the metal as a thin shield into a carrier metal cathode (Fig. 7b). The metal of the carrier should not radiate lines which lie close to the resonance lines of the analytical element and should not react chemically with it.

When the melting point of the metal is low it may be necessary to insert the metal into a cup-carrier cathode (Fig. 7c). Recently it has been claimed (15) that some advantage is to be derived from operating tin cathodes with the tin in a molten state. In these cases the carrier cathode must be such that the metal wets the sides "of the carrier." In addition the

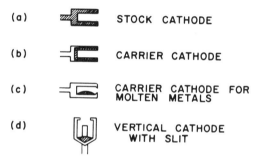

(a)   STOCK CATHODE

(b)   CARRIER CATHODE

(c)   CARRIER CATHODE FOR
       MOLTEN METALS

(d)   VERTICAL CATHODE
       WITH SLIT

Fig. 7. Types of cathodes used for lamps.

carrier cathode should have a cuplike form, i.e., with lips, to prevent molten metal from running out of the lamp. The authors were able to make very successful zinc lamps by constructing the cathode from brass and providing a zinc-metal reservoir (16). As the zinc sputtered from the brass it was replaced by diffusion from the zinc reservoir. With some low melting or mechanically weak metals, cathodes made from an alloy of one or more metals have been used.

Remarkably intense radiation from a lead hollow-cathode lamp was obtained by having a carrier cathode vertically positioned (17). A narrow vertical slit on the side of the cylinder enabled the negative glow to be observed from the side (Fig. 7d). Virtually no decrease in analytical sensitivity was observed when the lamp was operated at a high current and the lead was molten and sputtered profusely. When the lamp was observed under the same conditions from the open-ended side the resonance lines had lower intensity and gave much poorer analytical sensitivity.

2. Windows

Transmission of the resonant light is a prime requirement of the window. Most grades of borosilicate glass transmit visible and near ultraviolet light, but special materials are necessary for windows where far ultraviolet radiation is to be transmitted.

The ability to fuse the glass to the tube body is important in the manufacture of lamps as this can be a source for gas leakage. Should the window glass and the tube glass have different expansion properties, heating and cooling could cause cracks at the joint. Where it is possible to use windows with similar expansion and fusing properties as the tube glass, there is no difficulty. However, when ultraviolet light is to be transmitted, quartz or fused-silica windows are required. The expansion properties of these materials differ considerably from the normal tube glasses, and graded seals must be used to prevent uneven expansion. Some professional manufacturers attach the windows with epoxy glue, plasticized to allow the slight movements caused by expansion and contraction. The tube glass used for the hollow-cathode lamp should be chemically inert to the action of the metals sputtered onto them, e.g., alkalis and alkaline earths.

3. Lamp Conditioning

Certain special facilities are necessary to manufacture lamps. Ultra-cleanliness is a prime requisite. Even the slightest particle of organic matter, if included in the lamp, will give off sufficient impurity gases to end the life of the lamp rapidly. Out-gassing of all the metal and glass parts in a vacuum oven at a temperature above the release temperature is also necessary. After the electrodes have been introduced, and the lamp closed, it is sealed onto a vacuum system (13). The lamps are evacuated and heated again, preferably not with a flame as this introduces impurities which may diffuse into the lamp. Mercury pressure gauges and diffusion pumps should be avoided as mercury may "poison" the lamp. An induction furnace is most useful for heating and out-gassing the metal electrodes under high vacuum, and also for conditioning the getter.

Finally, it is obvious that the carrier gas used to fill the lamp should be ultrapure. It is useful to "getter" even an ultrapure grade of gases. After the lamp has been filled at the correct pressure it is sealed off, the getter is activated, and the lamp is ready for use.

It is possible to use continuous flushing of lamps in a laboratory, i.e., lamps which remain attached to the vacuum system with fresh carrier gas flushing through the lamp.

D. SPECIAL TYPES OF HOLLOW-CATHODE LAMPS

1. *Continuous Flushing of Demountable Lamps*

These types of lamps have been used extensively in the field of emission spectrographic analysis, and numerous designs have been published. They have also been applied for atomic absorption and atomic fluorescence purposes (*18–23*). Demountable lamps have the following advantages:

(a) Cathodes can be changed rapidly. This represents a saving in capital investment when a wide variety of lamps of different elements is used.
(b) According to Rossi and Amenetto (*19*) and Lang (*21*), higher intensities are obtainable because of less self-reversal. This is due to the cooling effect of the flushing gas.
(c) These lamps have a longer useful life than sealed tubes because loss of carrier gas by degassing is absent.

The reasons why demountable lamps are not more widely used than sealed lamps are:

(a) Sealed lamps are extremely simple and foolproof to use, even for untrained operators. On the other hand, a demountable lamp needs extra equipment to manufacture it, and its operation requires a higher degree of care and some training.
(b) Extra capital outlay in the form of vacuum systems and gas supplies is necessary. This offsets any saving obtained by not having to buy a variety of sealed lamps.
(c) Demountable lamps are probably less reliable than commercial lamps.
(d) More time is needed to change from one element to another.
(e) Continuous flushing lamps must be connected to large and cumbersome vacuum systems. Because of this, breakage of glass connections is common and the setup is clumsy.

The applicability of demountable lamps depends to a great extent on the particular circumstances in a specific laboratory. For further information on designs, readers are referred to the above-mentioned articles.

Information on suitable vacuum systems is given by Goleb and Brody (*24*), and Milazzo (*25*).

2. *High Intensity Lamps*

Sullivan and Walsh (*26*) developed a hollow-cathode lamp provided with supplementary electrodes similar to that shown in Fig. 8. When current is passed via these electrodes, a normal glow discharge takes place. The positive column of this supplementary discharge passes the mouth of the hollow cathode. Material sputtered from the cathode enters the

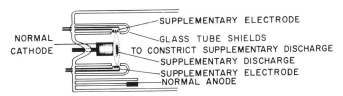

Fig. 8. Diagram of high intensity hollow-cathode lamp.

supplementary positive column and is excited. Davies (*27*) has suggested that the supplementary discharge reduces self-absorption broadening and that the light originates partly from the supplementary discharge and partly from the hollow-cathode discharge.

When oxide-coated electrodes, which have high electron emissivity, are used for the supplementary discharge, the discharge can be maintained with a relatively low voltage (about 40 V). A much higher current than is possible with the hollow-cathode discharge can be passed through the supplementary discharge. The low voltage and high currents result in a high intensity of the lower energy resonance lines.

This type of lamp is especially useful for atomic absorption analyses where an ionic or high energy line of an element interferes spectrally with the resonance line, e.g., nickel (*11*). The high intensity lamp may give up to 100 times greater intensity of the resonance line than normal open hollow-cathode lamps. This improves linearity of the working curve and may increase sensitivity.

High intensity lamps are excellent for atomic fluorescence purposes. Unfortunately, their lifetimes are not as good as normal hollow-cathode lamps. Supplementary oxide-coated filaments often burn out as a result of higher currents and overheating.

High intensity lamps are more difficult to make in the laboratory than normal hollow-cathode lamps. The filaments used for the supplementary discharge are made of nickel or tungsten wire coated with the "three-oxide" layer. Usually deposited as carbonates, the salts are converted to oxides by thermal processing. The three salts: barium-strontium-calcium carbonate are prepared in ultrapure form for the electron emissive surfaces, usually in vacuum tubes. The manufacturers of vacuum tubes have developed to a fine art the process of preparing "three-oxide" cathodes.

It is usual to start the supplementary discharge by passing a heating current through the filaments to make them more emissive. Once the filaments are hot the heating current may be reduced or switched off as the discharge energy is sufficient to keep the electrodes hot. Either dc or ac may be used to drive the supplementary discharge. In cases where ac is used it is necessary that both electrodes of the supplementary discharge be oxide-coated to ensure that the discharge will strike every time the current is changed.

It is debatable whether high intensity lamps hold any advantage for normal atomic absorption measurements except in cases where high intensity is needed, such as for selenium and arsenic, or where a spectrometer which has many mirror reflective surfaces is used, especially when the mirrors have deteriorated with age. It is claimed (14) that with the development of shielded cathodes the intensity of normal hollow-cathode lamps approaches that of high intensity lamps.

3. Multiple-Element Lamps

It has long been realized that a disadvantage of atomic absorption is the limitation of determining only one element at a time. To enable multiple-element analyses to be done, Massmann (28) and Butler and Strasheim (29) studied the use of multiple-element hollow-cathode lamps. Cathodes with rings of the various elements on the inside of a carrier cathode were made (see Fig. 8). As with some alloy lamps it was found that only certain elements could be combined. Where sputtering rates of the elements differed radically, the element with the lower sputtering rate would be covered by sputtered material from the other elements, and the intensity of spectral lines would decrease with time.

Sebens et al. (30) have reported success in the manufacture of multiple-element lamps by mixing the pure powders of elements and sintering them to form a cathode. High stability of emission for all the elements is reported.

The intensities of the different elements from multiple-element lamps are generally lower than from lamps where single-element cathodes are used. The more elements combined, the lower the individual intensities. Only those elements which have nonresonant and resonant spectral lines which do not overlap may be used in combination.

Multiple-element atomic absorption analysis does not appear to be used widely, and multiple-element lamps appear to be used mainly when a rapid change from one element to another is to be made on a single-channel monochromator. The authors know of no instance where multiple-element lamps have been used for multiple-element atomic fluorescence spectrometry.

VI. Vapor Discharge Lamps

Vapor discharge lamps are often called laboratory discharge lamps, arc lamps, Geissler lamps, line spectral lamps, and even by the name of the manufacturers, e.g., Wotan, Philips, or Osram lamps. Although designed primarily as sources for illumination, their high intensity and essentially arclike spectra make them well-suited as sources for atomic absorption and atomic fluorescence spectrometry. Dushman (31) has given an excellent treatise on the mechanism and manufacture of these lamps. Modern lamps, however, differ somewhat in construction from those described by Dushman.

A. MECHANISM OF THE DISCHARGE

While many manufacturers have different types of construction, most of them utilize the positive column type of discharge. It was mentioned earlier that the potential gradient of the positive column depends on the tube diameter, the dependence being approximately inversely proportional to the tube diameter. If the tube diameter is not too narrow the potential gradient is low, and far more secondary electrons are present than high energy primary electrons. Excitation of atoms of the metal vapor occurs by collisions of secondary electrons with atoms. The reasons for this are as follows:

Primary electrons released from the oxide-coated cathode are strongly accelerated through the Crookes dark space in the same manner as for any other type of glow discharge. The electrons have sufficient energy to ionize carrier gas atoms. The secondary electrons released as a result of these collisions do not have sufficient energy to further ionize carrier gas atoms, but can excite metal atoms with low excitation potentials. As most of the secondary electrons exist in the positive column, the most intense spectra are obtained from this region. The current density through the tube and the density of metal atoms present are the main contributing factors to the intensity of radiation. Unlike the hollow cathode, positive gas ions do not cause profuse sputtering from the cathode because of the high thermionic electron emissivity. Positive ions attracted to the cathode rather remove the space charge built up near the cathode by electrons.

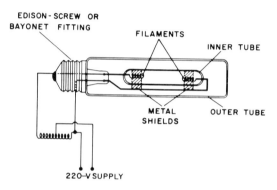

Fig. 9. Diagram of vapor discharge lamp.

Those elements which have relatively low excitation potentials and which are easy to vaporize are better suited for vapor discharge lamps, e.g., the alkali metals, mercury, cadmium, and zinc. When the lamps are made of these elements the lowest energy spectra, or resonant radiation are the most intense. Indeed, higher energy lines and ionic lines are generally so weak that they are barely seen except when their excitation energy is very low (e.g., cesium).

Figure 9 shows a schematic diagram of a discharge tube. The positive column occurs in the central portion of the tube. Electrodes are sealed into the inner discharge tube. An outer tube serves to provide heat insulation to prevent condensation of the metal on the inner tube walls. Earlier types of vapor discharge tubes had electrodes with the filament heated by a supplementary current, but modern construction and the development of the high electron emissivity "tri-oxide" cathodes enable lamps to be made without supplementary heating.

When the voltage appears across the tube after switching on, a certain degree of ionization is present in the carrier gas to enable the discharge to start. This takes place more easily when the voltage across the electrodes is relatively high. Considerably higher currents than for hollow-cathode lamps are passed through vapor discharge lamps. The heating process of the current heats up the walls and the cathodes. The condensed metal which might be in contact with the cathode or on the walls slowly evaporates and the vapor enters the discharge column. The process of evaporation takes place fairly slowly and the lamps are characterized by strong emission of the carrier gas spectrum until the metal has evaporated sufficiently. (This effect is often seen in sodium street lamps soon after switching on, when they appear a deep red, due to the excitation of the neon carrier gas.)

B. Use of Vapor Discharge Lamps

The electrical power supplies for vapor discharge lamps are simple compared to those required for hollow-cathode lamps. The discharge lamps could be connected to an ordinary 220-V supply with some means to limit the current, as the burning voltage may be as low as 30 or 40 V. However, ignition of the discharge may present difficulties, and a simple low-power-factor autotransformer with an open circuit voltage of between 400–500 V is usually used (see Fig. 9). The low-power-factor transformer limits the current automatically so as not to exceed about 900 mA.

It has been mentioned in the literature that it is preferable to reduce the current through the lamps to gain sensitivity. This may be done by means of a series resistance. Reducing the current reduces the temperature, and thus the broadening of lines due to Doppler (temperature) effects. Care should be taken, however, not to reduce the current through the lamps too much, as it has been the experience of the authors that too great a reduction of current often leads to self-absorption broadening and instability of the discharge. It has also been found that the radiation intensity of the lamps may change due to the condensation of metal vapor on the too-cool walls of the discharge tube.

Since hollow-cathode lamps of the alkali metals have become available, less use is made of vapor discharge lamps for atomic absorption purposes. However, certain of the alkali metals have their resonance lines in the red or near infrared region of the spectrum, and many monochromators have gratings which are not blazed for this region and photomultipliers which are relatively insensitive in these regions. In these cases more stable readings may be obtained because of the lower amplifier gain settings but a loss of sensitivity is found when the high intensity vapor discharge lamps are used (32).

Because of their high intensity, vapor discharge lamps are well suited for use in atomic fluorescence spectrometry. It is found, however, that the self-absorption which is inherently present in these lamps leads to a decrease in fluorescence sensitivity when compared to high intensity hollow-cathode and electrodeless discharge sources (33).

VII. Electrodeless Discharge Sources

Electrodeless tubes were originally described by Jackson (34) and used by Meggers (35) for high resolution spectral studies. More recently they have found increasing use as sources for atomic absorption and especially for atomic fluorescence spectrometry. Their high intensity, freedom from self-absorption, and narrow spectral lines make them ideally suited for these applications. In addition, they can be made for a wide variety of elements. It has also been reported that they are easy to manufacture and easy to use.

Several papers have described the manufacture and characteristics of these tubes (33, 36, 37), and more information is appearing continuously about them.

The electrodeless discharge tube is a thin quartz or glass tube into which is sealed a small amount of the element under a low pressure of inert gas. The tube is placed in a radio-frequency or microwave field, and excitation of the metal vapor takes place. Coupling of the electrical discharge with the vapor is usually done via suitable coils or wave guides.

A. Manufacture of the Lamps

The thin translucent tube used to contain the metal is usually made of quartz because of the high temperatures that are evolved in the cavity and also to enable the transmission of ultraviolet light. The diameter is usually between 5 and 12 mm and the length of the tube up to 7 cm. Mansfield et al. (36) found that the length of the tube and the weight or volume of material within the tube were not critical for most elements. They also found that an inside diameter of less than 9 mm was better than larger diameters for a stable discharge. Tubes with a diameter much less than 5 mm tended to be excessively noisy and had short lifetimes, although they had higher intensities than larger diameter tubes.

The material in the tube must have a relatively low vapor pressure of about 1 Torr at 200–400°C (26) to be vaporized easily. It is for this reason that metal halides, usually iodides, are used for metals with high vapor pressures. Dagnall and West (33) have reported that it is useful to form the halide in situ by the additions of a few milligrams of metal and iodine such that the metal is present slightly in excess of the iodine. Mansfield et al. (36) reported that the use of mercury amalgam lamps, i.e., where the material in the lamp consists of the metal halide together with a certain amount of mercury, were much noisier although more intense than the metal or metal iodide lamps.

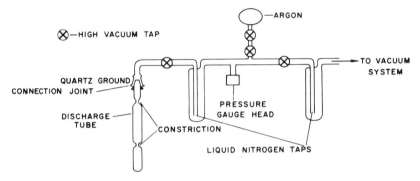

Fig. 10. Diagram of electrodeless discharge tube attached to the vacuum line

The tubes are connected to a vacuum system and thoroughly outgassed before the metal is added. As with any vacuum technique, the outgassing process and cleanliness of the materials is of the utmost importance. Once the tubes have been properly outgassed, the material is introduced either directly as the metal, or as a piece of metal wire together with a certain amount of iodine, or else introduced into the bulb of the tube by sublimation. Figure 10 shows a diagram of a tube for introducing low melting point metals into the discharge tube. (Several workers claim that an excess of the *metal* is an advantage.) As most vacuum systems are made from borosilicate glass, a graded glass tube is necessary to enable the quartz tube to be sealed onto the system, or the connection can be made by a connecting joint similar to that shown in Fig. 10.

After the material has been introduced into the discharge tube and it has been thoroughly evacuated and outgassed, a small amount of argon is introduced at a pressure of 1–3 Torr. Higher intensities of radiation are obtained for lower argon pressures, but with these lower pressures the lifetimes of the lamps are reduced. After the lamps have been detached from the vacuum system and sealed, they usually require a "breaking-in" period, i.e., a period where they are operated at high energy.

B. THE POWER SOURCE

The frequency range which has been used for exciting the electrodeless discharges is from 10 MHz to 3000 MHz. While lower frequencies may be used for the excitation of the alkali metals with low vapor pressures, it is more common to use higher frequencies in the microwave region for most metals. Medical diathermy units operating at a frequency of about 2450 MHz with powers up to 150 W have been found to be useful as microwave excitation sources. They are relatively inexpensive, easy to operate, supply a constant source of microwaves, and a range of discharge cavities are offered.

The efficiency with which the electrodeless discharge tube operates depends to a large extent on the proper choice of the discharge cavity. Energy from the microwave source must (i) transfer energy to the metal or salt to cause vaporization and (ii) excite the vapor to radiate characteristic light. It is for this reason that much attention has been given to the design of suitable cavities. The cavity is designed to resonate and to reflect as much of the power into the discharge tube as possible. When the resonant frequency of the cavity containing the gas tube is tuned to that of the source, the impedance is matched and little power is reflected back to the microwave unit. If too much power is reflected back this may cause damage to the magnetron. Reflected power meters are therefore a necessary accessory. Cavities which are relatively independent of the gas pressure in the discharge tube and the material in the gas discharge tube are preferred. The conditions for resonance may change after the start of the discharge, and it is therefore necessary for a tuning facility to be on the cavity. A good deal of heat is also generated and it is advisable to have an aperture for passing clean air over the surface of the tube to keep it reasonably cool. An A type cavity, as used by Mansfield et al. (36), is shown in Fig. 11. In order to keep the temperature of the discharge tube high to prevent condensation and erratic discharges to various portions of the tube when certain elements are excited, the tube may be inserted either into a container insulated with quartz wool (Fig. 11) or be in a vacuum container.

The $\frac{3}{4}$ wave (cylindrical) and $\frac{1}{4}$ wave (Evenson) cavities (Fig. 12) have been quite successful in exciting gas discharges in both static and flowing systems. The cylindrical cavity (Fig. 12A), while not as efficient as a properly adjusted Evenson cavity, obviates the need for repeated tuning adjustment; only a single adjustment is provided. In general, the cavity must be operated at powers 25% greater than the Evenson cavity to achieve similar results. The cavity is constructed so that the lamp is

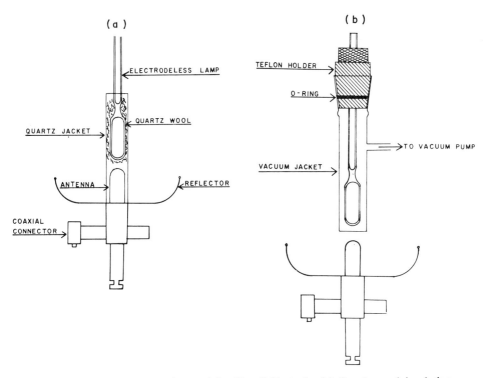

Fig. 11. Diagram of cavity used by Mansfield et al.: (a) Quartz wool insulation [Reprinted from Ref. (*36*) by courtesy of Pergamon Press and authors]; (b) Vacuum insulation [Reprinted from Ref. (*37*) by courtesy of American Chemical Society and authors].

protected from drafts in the laboratory and so operates at essentially a constant temperature. Two adjustments are provided for the Evenson cavity (Fig. 12B and C) to properly match the impedance of the discharge to that of the generator, which improves the efficiency of the system. Lamps containing the more difficult elements such as Al, Si, and Ti are easily excited by the Evenson cavity. For either of the above cavities it is

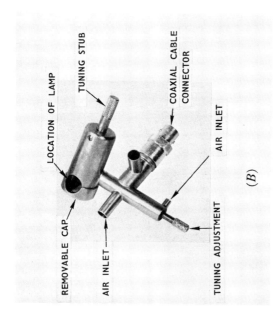

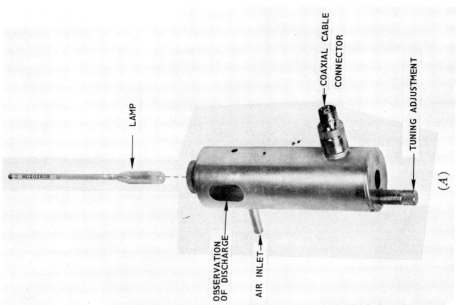

Fig. 12. Cavities for electrodeless discharge lamps: (A) $\frac{3}{4}$ wave (cylindrical) cavity; (B) $\frac{1}{4}$ wave cavity (Courtesy of Electro-Medical Supplies Ltd.); (C) $\frac{1}{4}$ wave (Evenson) cavity; (D) $\frac{1}{4}$ wave modified (Evenson) cavity. (Courtesy of Ophothos Instrument Co.)

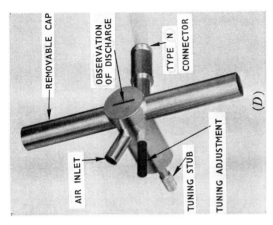

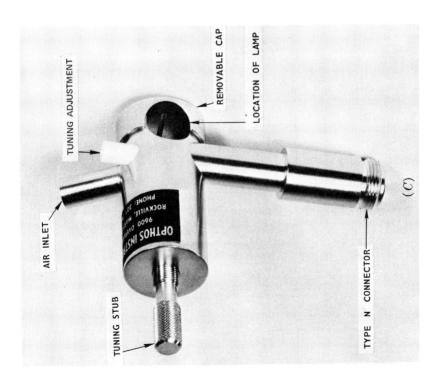

Fig. 2-12) (cont.)

strongly recommended that one use a bi-directional power meter to monitor the forward and reflected power.

An improvement in design of the Evenson cavity is shown in Fig. 12D (*38*). The modified cavity eliminates arcing between the tuning stub and the coupler and permits adjustment of the reflected power to less than 1 W at rate transmitter output of 100 W. The virtue of low values of reflected power is that the losses in the cable between the transmitter and cavity are minimized, tuning of system is easier, and peak voltages are reduced.

Electrodeless discharge tubes offer definite advantages to the analytical spectroscopist who wishes to do atomic fluorescence spectrometry. It has been reported by Dagnall and West (*33*) that electrodeless discharge tubes give better stability and have shorter warm-up periods than do hollow-cathode lamps. The possibilities of making multiple-element lamps are being investigated (*39*). Electrodeless discharge tube may well solve this problem in atomic absorption and atomic fluorescence spectrometry.

VIII. Other Spectral Sources

All line sources possess one serious disadvantage, namely, a different source is required when different elements are to be measured by atomic absorption or atomic fluorescence spectrometry. Multiple-element hollow-cathode and electrodeless discharge lamps have partially solved this problem, but the analyst is still severely limited to those elements put into the lamps, i.e., lamps must be specially made for a purpose. If a wide program of elements is to be measured, a large number of lamps is required, which can be expensive. In addition, the use of line sources limits the methods to quantitative methods, namely, to determine the concentration of *known* elements. Should it be required to determine the constituent elements of the sample, the analyst must resort to emission spectrographic or x-ray techniques. Several groups have studied the use of other sources with a view to overcoming these difficulties.

A. Continuous Light Sources

The use of continuous light sources was suggested by Walsh (*1*) in the now famous first paper on atomic absorption. He did not recommend continuous sources because of the obvious difficulties associated with

their use, namely: (1) the high resolution required to isolate the absorption spectral line in a continuous spectrum. Due to the narrowness (0.01 to
0.03 Å) of the resonance absorption lines, a resolution of better than
500,000 would be necessary. (2) The diffculty of obtaining sufficient
energy in that spectral region isolated by the monochromator to enable
sensible low noise measurements to be made.

The group studying the use of continuous sources have found that in
certain cases useful absorbances could be measured with medium dispersion monochromators (40). Fassel et al. (41) have reported improved
sensitivities for certain elements including the rare earths. Sensitivities
similar to those obtained by atomic absorption with hollow-cathode lamps
were obtained for many of the alkaline earths and transition elements.

The surprisingly high sensitivity of these results confirms that the simple
model of the absorption line profile as suggested by Fassel may not be
correct, especially for those elements which have complex spectra and
hyperfine structures. Allan (42) recently showed the importance of the
hyperfine structure of the absorption line. It is interesting to note that the
high sensitivities obtained with a continuous source were with those
elements which have complex spectra and whose resonance lines fall
mainly in the visible regions of the spectrum. With a multitude of absorbing
components of a line it is probable that the absorption is spread over a
relatively larger spectral region, hence the better absorption when a
relatively wide bandwidth is covered from a continuous spectrum. The fact
that an iron spectrum gives fairly good absorption for many elements (43)
confirms this.

Tungsten filaments radiate continuous spectra. They are well suited for
measurements in the red and visible regions, but due to the blackbody
radiation peak being at longer wavelengths, intensity falls off rapidly at
shorter wavelengths. Quartz halide lamps used by the authors have enabled
measurements as far down as 3000 Å. For lower wavelengths strong
ultraviolet radiators have been preferred. Hydrogen-arc and xenon
lamps have also been found to be satisfactory (44). Xenon- and hydrogen-
arc lamps do not show blackbody radiation characteristics, as the radiation is from strongly broadened lines and "Bremsstrahlung" continua.
The authors have found many xenon arcs were prone to be unstable.

Continuous sources offer advantages for certain purposes. It is necessary,
however, that they radiate very strongly to obtain sufficient energy in the
small spectral region selected to enable measurements. It is usual to scan
the region of the absorption spectral line to measure an absorption peak.
An atomic absorption instrument which makes use of continuous sources

and has a very high resolution etalon (interferometric) spectrometer has recently appeared on the market.

B. FLAMES

A combustion flame radiating the atomic spectra of the analytical element is a possible source for atomic absorption. The use of flames has been reported (45, 46) in atomic absorption. The radiative intensity is too low for atomic fluorescence measurements, however.

Manning and Slavin (47) and Alkemade and Milatz (48) have used a flame source for lithium, and Skogerboe and Woodriff (49) have used a oxygen–hydrogen flame as a source for determining some rare earths. Flames generally exhibit self-absorption, and this limits the sensitivity of atomic absorption. In addition, even the hottest combustion flames are not capable of exciting atoms with medium and higher excitation potentials to radiate strongly enough for absorption measurements. Rann (46) has evaluated, theoretically and experimentally, the flame as a source and concludes that it compares poorly with a hollow-cathode lamp, both with regard to sensitivity and stability. It does have the advantage that several elements can be excited simultaneously.

A plasma-arc flame has also been used as a source for atomic absorption (50). While the plasma temperature is high (20,000°K), suprisingly good absorption was obtained. Alkemade (2) has confirmed that this is possible.

C. SPARKS

Strasheim and Human (51) have reported the use of a spark as a source for atomic absorption. The spark is known as an intense source of light but has broad spectral lines. It is known, however, that broadening and background radiation occur mainly in the initial portion of the discharge, and that the latter portion or afterglow of individual discharges radiates relatively narrow spectral lines. A rotating disk coupled to a high precision synchronous spark enabled the initial portion of the discharge to be rejected and only the latter portion of the discharge to be measured. When the light was passed through an atomic absorption flame, absorption occurred. Surprisingly good sensitivity was reported, but the stability of the source was poor in comparison with a hollow-cathode lamp. Multiple-element analyses were made possible by sparking solutions of the various elements.

D. SELECTIVE MODULATORS

Sullivan and Walsh (52) have reported the development of a new type of source. They make use of a hollow-cathode lamp and pass the light through an atomic cloud provided by another set of electrodes. Atomic absorption occurs in this atomic cloud so that the resonance lines from the hollow cathode are self-absorbed. The discharge causing the atomic cloud is switched on and off at fixed frequency so that the intensity of the

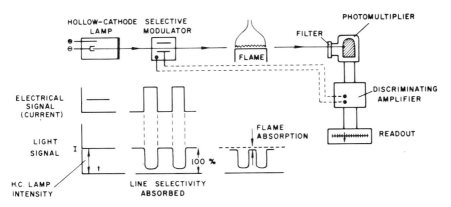

Fig. 13. Schematic diagram of selective modulation system.

resonance lines will be modulated at that frequency. If the detector-amplifier is locked to that frequency, it is possible to differentiate resonance lines from nonresonance lines without the use of a spectrometer. A schematic diagram of the system is given in Fig. 13.

A further development of this system is the use of solar-blind photo-multipliers (53). A special technique for pulsing the current through a normal hollow-cathode lamp while the normal discharge is taking place and so obtain a similar effect to that of a separate modulator has been published by Lowe (54).

E. NARROW LINE SOURCE

Van Gelder (55) has developed a novel high intensity narrow line source. This source is shown schematically in Fig. 14. A positive column discharge is passed through the center of a ring electrode. This electrode is connected electrically to the cathode of the discharge in such a way that it is negative with respect to the positive column discharge. Ions in the positive column are attracted and bombard the electrode surface causing sputtering. By restricting the positive column before and after the electrodes by means of a constriction in the glass tube, a high current density is created immediately in front and behind the supplementary electrode. Excitation in the restricted plasma by secondary electrons takes place on

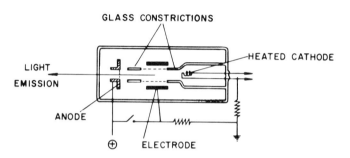

Fig. 14. Van Gelder high intensity source.

the tube axis where the atomic density is low but electron density is high. The result is extremely intense but narrow arc-like spectral lines.

IX. Conclusion

The development of sources for atomic absorption and atomic fluorescence spectrometry has taken place very rapidly during the last few years. It has not been possible to consider all the sources that have been described in the literature. The sources in use today are generally satisfactory for their purpose, but there will certainly be further developments. Perhaps the problems that exist today with regard to intensity, stability, long life, and the ability to radiate the light of several elements simultaneously will be solved.

One may well see the emergence of laser-type sources which radiate characteristic-type spectra (56). It is felt that sources such as electrodeless discharge lamps and selective modulators show considerable promise and may well supersede other sources for atomic absorption and atomic fluorescence spectrometry in the future.

REFERENCES

1. A. Walsh, *Spectrochim. Acta*, **7**, 108 (1955).
2. C. T. J. Alkemade, *Appl. Opt.*, **7**, 1261 (1968).
3. J. D. Winefordner and T. J. Vickers, *Anal. Chem.*, **36**, 161 (1964).
4. F. Llewellyn-Jones, *The Glow Discharge*, Methuen, London, 1966.
5. S. C. Brown, *Introduction to Electrical Discharge in Gases*, Wiley, New York, 1966.
6. L. B. Loeb, *Fundamental Processes of Electrical Discharges in Gases*, Wiley, New York, 1939.
7. F. Paschen, *Ann. Physik*, **50**, 901 (1916).
8. H. Schüler, *Z. Physik*, **35**, 323 (1926).
9. D. J. Sturges and H. J. Oskam, *J. Appl. Phys.*, **35**, 2887 (1964).
10. K. B. Mitchell, *J. Opt. Soc. Am.*, **51**, 846 (1961).
11. W. Slavin, *Atomic Absorption Spectroscopy*, Interscience-Wiley, New York, 1968.
12. A. Strasheim and L. R. P. Butler, *Appl. Spectry.*, **16**, 109 (1962).
13. W. G. Jones and A. Walsh, *Spectrochim. Acta*, **16**, 249 (1961).
14. J. C. Burger, Westinghouse Corp. Engineering Memo ETD 6702, June 1967.
15. J. Vollmer, *At. Abs. Newsletter*, **5**, 35 (1966).
16. L. R. P. Butler, *S. African Ind. Chemist*, **15**, 162 (1961).
17. P. B. Zeeman and L. R. P. Butler, *Appl. Spectry.*, **16**, 120 (1962).
18. R. E. Popham and W. G. Schrenk, *Appl. Spectry.*, **22**, 192 (1968).
19. G. Rossi and N. Omenetto, *Appl. Spectry.*, **21**, 329 (1967).
20. G. I. Goodfellow, *Appl. Spectry.*, **21**, 39 (1967).
21. W. Lang, *Z. Instrumentenk.*, **75**, 216 (1967).
22. S. R. Koirtyohann and C. Feldman, *Developments in Applied Spectroscopy* (J. Forette, ed.), Vol. 3, Plenum Press, New York, 1964.
23. J. I. Dinnin and A. W. Helz, *Anal. Chem.*, **39**, 1489 (1967).
24. J. A. Goleb and J. K. Brody, *Appl. Spectry.*, **15**, 166 (1961).
25. G. Milazzo, *Appl. Spectry.*, **21**, 185 (1967).
26. J. V. Sullivan and A. Walsh, *Spectrochim. Acta*, **21**, 721 (1965).
27. D. K. Davies, *J. Appl. Phys.*, **38**, 4713 (1967).
28. H. Massmann, *Z. Instrumentenk.*, **71**, 225 (1963).
29. L. R. P. Butler and A. Strasheim, *Spectrochim. Acta*, **21**, 1207 (1965).
30. C. Sebens, J. Vollmer, and W. Slavin, *At. Abs. Newsletter*, **2**, 165 (1964).
31. S. Dushman, *J. Opt. Soc. Am.*, **27**, 1 (1937).
32. B. J. Russell, J. P. Shelton, and A. Walsh, *Spectrochim. Acta*, **8**, 317 (1957).
33. R. M. Dagnall and T. S. West, *Appl. Opt.*, **7**, 1287 (1968).
34. D. A. Jackson, *Proc. Roy. Soc. (London)*, **A121**, 432 (1928).
35. W. F. Meggers, *J. Opt. Soc. Am.*, **38**, 7 (1948).
36. J. M. Mansfield, M. P. Bratzel, H. O. Norgordon, K. Zacha, D. O. Knapp, and J. D. Winefordner, *Spectrochim. Acta*, **23B**, 389 (1968).
37. C. H. Corliss, W. R. Bozman, and F. O. Westfall, *J. Opt. Soc. Am.*, **43**, 398 (1953).
38. B. McCarroll, *Rev. Sci. Instr.*, **41**, 279 (1970).
39. T. S. West, personal communication.
40. J. H. Gibson, W. E. L. Grossman, and W. D. Cooke, *Anal. Chem.*, **35**, 266 (1963).
41. V. A. Fassel, V. G. Mossotti, W. E. L. Grossman, and R. N. Kniseley, *Spectrochim. Acta*, **22**, 347 (1966).
42. J. E. Allan, *Spectrochim. Acta*, **24B**, 13 (1969).
43. C. W. Frank, W. G. Schrenk, and C. E. Meloan, *Anal. Chem.*, **38**, 1005 (1966).

44. C. Veillon, J. M. Mansfield, M. L. Parsons, and J. D. Winefordner, *Anal. Chem.*, **38,** 204 (1966).

45. C. S. Rann, *Spectrochim. Acta,* **23B,** 245 (1968).

46. C. S. Rann, Ph.D. Thesis, Australian National University, Canberra, 1967.

47. D. C. Manning and W. Slavin, *At. Abs. Newsletter,* **1,** 39 (1962).

48. C. T. J. Alkemade and J. M. W. Milatz, *J. Opt. Soc. Am.,* **45,** 583 (1955).

49. R. K. Skogerboe and R. Woodriff, *Anal. Chem.,* **35,** 1977 (1963).

50. H. G. C. Human, L. R. P. Butler, and A. Strasheim, *Analyst,* **94,** 81 (1969).

51. A. Strasheim and H. G. C. Human, *Spectrochim. Acta,* **23B,** 265 (1968).

52. J. V. Sullivan and A. Walsh, *Appl. Opt.,* **7,** 1271 (1968).

53. P. L. Larkins, R. M. Lowe, J. V. Sullivan, and A. Walsh, *Spectrochim. Acta,* **24B,** 187 (1969).

54. R. M. Lowe, *Spectrochim. Acta,* **24B,** 191 (1969).

55. Z. Van Gelder, *Appl. Spectry.,* **22,** 581 (1968).

56. T. F. Trost and W. B. Johnson, NASA Tech. Rept. A46.

3 Nebulizers and Burners

Roland Herrmann

DEPARTMENT OF MEDICAL PHYSICS
UNIVERSITY OF GIESSEN
GIESSEN, WEST GERMANY

I. Introduction

The most important component in flame spectrometry, whether emission or absorption, is the nebulizer and burner system. The function of the nebulizer is to guarantee a uniform and efficient supply of sample to the flame in as simple a manner as possible. In practice, the sample, usually in liquid form, is converted to a gas-liquid aerosol in the nebulization step and the aerosol then directed into the flame. The function of the burner is to assure a flame sufficiently stable and reproducible in its physical-chemical properties so that together the nebulizer and burner ensure uniform material transfer with the least signal fluctuation. In addition, the burner must produce a flame sufficiently energetic to convert the analyte to atomic vapor and, in emission, to excite the characteristic radiation of each test element.

Originally the design and theory of operation for nebulizers and burners were concerned only with the emission method; however, they are also valid for atomic absorption and atomic fluorescence methods. Historical aspects are treated in references (1–5).

One distinguishes between the nebulizer, the spray chamber, the burner, and the transfer path taken by the aerosol. Nebulization is generally carried out with the support gas. When the liquid is sprayed first into a mixing chamber and then the droplets mixed with the combustion gas, the arrangement is denoted *indirect nebulization*. In this configuration a spray chamber lies between the nebulizer and burner with all parts connected. In the second category the nebulizer-burner unit is combined. The burner tip is the point at which nebulization takes place and where the combustion gas is admixed. The sample aerosol passes directly into the flame without a transfer step. These units are called direct sprayer-burners. Lastly, there is a transitional category wherein the transfer path is finite but extremely short.

II. Methods for Sample Introduction

The production of aerosols from liquid or solid samples is an essential process in most analytical applications of flame emission and atomic absorption spectrometry. Although an aerosol may be produced by a variety of techniques, pneumatic devices are the most widely used.

A. Pneumatic Nebulization

The construction of a typical pneumatic nebulizer is shown in Fig. 1. Compressed air or oxygen streams out of the nebulizer aperture around the solution capillary at approximately the speed of sound. According to

Bernouilli's principle, a back pressure of about 25 cm of Hg is created which draws the test solution into and up the capillary. The emerging liquid stream is broken up into droplets by the high velocity gas stream. In addition the shock waves generated by the edges of the nozzle in the supersonic gas flow may be expected to exert a disruptive effect on passing droplets, perhaps greatest on the larger droplets. An aerosol is formed with droplets of test solution surrounded by air or oxygen.

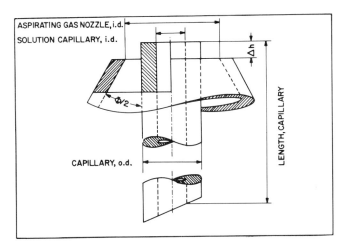

Fig. 1. Schematic of a typical pneumatic nebulizer. $\phi/2$ is convergence angle of aspirating gas stream; Δh is extension of capillary beyond the aspirating nozzle.

In the nebulization step, work must be performed to overcome the surface tension and viscosity of the liquid. The necessary energy is supplied from the kinetic energy of the aspirating gas stream. With higher energy transfer to the nebulizer, fewer larger droplets are produced and the energy is transferred to the test liquid (2, 6–8).

Surface tension and viscosity are important solution properties and must be maintained as nearly identical as possible in samples and standards. This can be approached quite simply by diluting the test solution unless one is working near detection limits. Concentrated solutions should be diluted to avoid encrustation of salts on the nebulizer and burner, and to minimize any effect of foreign bodies when handling heterogeneous systems such as colloids and solutions high in protein content. Although little is gained through reducing the surface tension by addition of surface-active agents, as discussed in Chapter 10 of Volume 1, addition of wetting agents is recommended for biological solutions to keep the inner capillary wall free from adsorbed materials. Any wetting agent employed must be free from metal impurities, and similar amounts must be added to standards and samples alike.

From the Hagen-Poiseuille law the quantity of liquid flowing through the capillary is proportional to the differential pressure at the solution capillary and inversely proportional to the viscosity of the solution sprayed. In general, the greater the negative pressure, the better the nebulization. The amount of solution reaching the flame will decrease with increasing viscosity. However, the effect of viscosity will be slight, as discussed in Chapter 10 of Volume 1. Height between the tip of the nebulizer and the solution level is also involved. Large variations in the level of the sample solution must be avoided to maintain the hydraulic head reasonably constant. A uniform time pattern should be maintained in placing solutions under the capillary and removing. The nebulization interval should be kept constant, as well as the time period intervening between aspirating solutions. An automatic sample changer is ideal in this respect since it operates on a programmed cycle. When working with volatile solvents, special precautions are necessary (2, 9). Several methods have been proposed to deliver a constant quantity of solution to the nebulizer such as motor-driven syringes or peristaltic pumps (10–12).

A nebulizer is rated in terms of its efficiency, i.e., the portion of test solution which is converted to solid residue and which actually vaporizes in the flame to appear in atomic form per unit time, as compared to the total consumption of test solution. In conjunction with a mixing chamber, nebulization efficiency can be appreciably improved by placing an impact surface or sphere (brise-jet) at a distance of a few millimeters from the nebulizer tip and in the jet of liquid emerging from the solution capillary (2, 3). Since the droplet impinges against the surface of the bead, droplet size is thereby reduced. A premix counter-flow jet operates similarly (13); it directs a stream of premixed oxidant and fuel at the output stream of the nebulizer in the spray chamber. The face of the gas stream disrupts the solvent droplets to produce a finer aerosol. The jet can be rotated in or out of the liquid stream, as well as moved back and forth to optimize nebulization for different solutions (Fig. 2).

Sample aspiration rate can be monitored in several ways. Sample consumption can be measured with a stop watch. The time interval between placement of the capillary into the sample and the first appearance of a flame signal can be measured. Quantity of aerosol and its fineness can be observed visually; also see Chapter 1 (Section IV.D).

The optimal operating pressure of the aspirating gas must be ascertained experimentally. It is a function of the test element, the solvent, the nebulizing gas, the matrix composition, and the physical-chemical properties of the solution (*1*). The combustible gas supply also plays an essential role. The optimum pressure (and flow rate) of aspirating gas depends on whether one wishes (1) to optimize in terms of good reproducibility at higher concentrations, (2) to strive for useful deflections at lower concentrations, (3) to minimize flame background contributions,

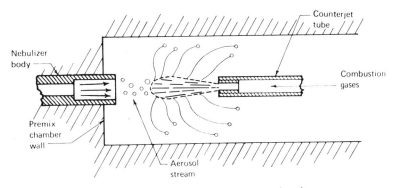

Fig. 2. Counter-flow jet within the spray chamber.

especially during emission measurements near the detection limits, or (4) to increase the signal-to-noise ratio. Optimum conditions will usually differ for different spectrometers, for individual elements, and whether an emission or absorption method is involved. Unfortunately, the optimal operating pressure for nebulization and transfer efficiency is not necessarily the best for the subsequent signal generation in the flame gases. The flame temperature will also change when the nebulizing gas supply is altered. Thus, one should optimize the nebulizer efficiency only in conjunction with other parameters (*1, 2, 10, 14–20*).

Once the optimal operating gas pressures have been ascertained, they must be adequately regulated. Two-stage pressure relief valves are recommended. To guard against possible obstructions in the nebulizer, gas flows should be monitored by rotameters or flow meters (*1–3*).

The ideal aerosol should be of high sample mass per unit volume of gas, homogeneous, and of small droplet size. Usually, these requirements are incompatible with each other. The pneumatic nebulizer produces a heterogeneous aerosol which contains droplets of widely varying diameters (1 to 70 μm), but in which the majority of droplets have a diameter of 5 to 10 μm, while most of the sample volume is contained in droplets with a diameter of 20 μm or greater. At very low sample aspiration rates, less than 0.1 ml/min, nearly complete desolvation can occur before the aerosol reaches the flame. Hence, atomization will be from a heterogeneous dust cloud (*21*). Droplet aggregation by collision will be infrequent. With higher sample aspiration rates, some droplets may not be completely desolvated before entering the flame area in the optical path of the spectrometer or will be lost by deposition on the chamber walls if a mixing chamber is involved in the transfer path. Vaporization and atomization of solids in the flame will be delayed from the clotlets remaining from the larger droplets. The size of the dry particles in the flame is proportional to that of the droplets from which they are formed; larger droplets lead to larger solid particles which will require more time to complete the vaporization, dissociation, and atomization steps. With very large droplets or samples forming refractory compounds, they may never be achieved. Also scattering of primary light by undesolvated droplets in the flame would be included as part of the analyte signal in atomic flame fluorescence, and would lead to an increase in absorbance due to scattered light in atomic absorption. The noise level would also increase.

B. Construction Features of Pneumatic Nebulizers

An aerosol composed of finer droplets can be attained by using higher energies, aside from the use of special solutions. Since the energy for nebulization must be taken from the kinetic energy of the aspirating gas, high gas velocities increase the kinetic energy available. In normal nebulization the upper limit is the speed of sound. Energy, expressed at $mv^2/2$, cannot be increased indefinitely merely by increasing the velocity v but only by increasing the mass m of the aspirating gas through greater compression. Since the latter quantity varies linearly and not as the square of gas velocity, nebulization efficiency is not significantly increased when the aspirating gas differential pressure is raised above 1 kg/cm² (or 1 atm). For this reason, and because of the larger amount of liquid introduced, the curves in Fig. 3 become flatter at elevated pressures. On the other hand, at pressures substantially less than 1 kg/cm², the efficiency deteriorates significantly. Good nebulizers should be built to operate between 0.7 and 1 kg/cm². At high pressures a great deal of aspirating gas will flow through normal size nozzles. Too high a flow velocity requires more

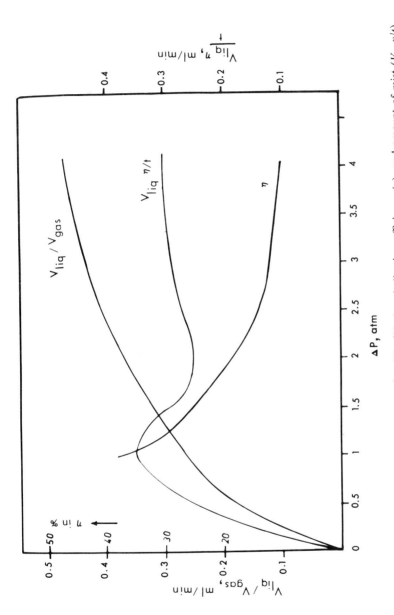

Fig. 3. Dependence of aerosol concentration (V_{liq}/V_{gas}), nebulization efficiency (η), and amount of mist ($V_{liq}\eta/t$) aspirating gas pressure with concentric nebulizers. [Reprinted from (55) by courtesy of Archiv für das Eisenhütten-wesen.]

combustible gas to maintain a desired mixture strength. The flame
dimensions would become larger, thus requiring more test solution.
Moreover, a substantial part of the flame would lie outside the optical
path and would not be utilized in the actual measurement step. Consequently, small nozzle openings in the nebulizer are necessary.

1. *Angle Nebulizer*

In the angle nebulizer (Fig. 4) the liquid and aspirating gas pass through
two perpendicular tubes. Utilization of the aspirating gas stream for

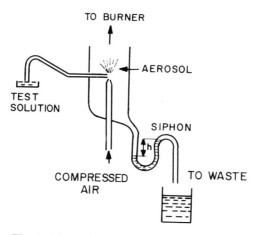

Fig. 4. Schematic diagram of an angle nebulizer.

nebulization is not very efficient in this design. A large portion of the gas
stream sweeps unused past the orifice of the solution capillary. Rate of
solution uptake and droplet size depend critically upon the relative
position of the two tubes. Obstructions frequently narrow the orifices at
the nebulization site or stop aspiration altogether. The only advantage
of this design lies in the ease of fabrication since the nebulizer is usually
blown from glass or constructed from two hypodermic needles.

2. *Concentric (Split-Ring) Nebulizer*

In this type of nebulizer a central liquid intake capillary lies within a
concentric (split-ring) nozzle for the perfusion of aspirating gas (Fig. 5).
Compared with the angle nebulizer, a concentric nebulizer produces

smaller drops with the same aspirating gas pressure, and gas and liquid flow.

As shown in the upper row of Fig. 5, the convergence angle of the aspirating gas nozzle to the central solution capillary can be varied. When

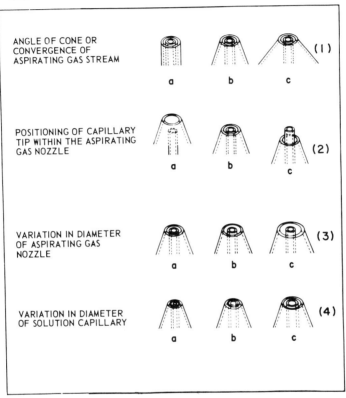

Fig. 5. Concentric (split-ring) nebulizer. Row 1, angle of convergence of aspirating gas stream; row 2, positioning of capillary solution tip within the nozzle of the aspirating gas; row 3, variation in diameter of aspirating gas nozzle; and row 4, variation in diameter of the solution capillary.

the angle of attack is zero, the capillary and aspirating gas annuli are parallel. Fabrication is simple. However, the relatively long annulus presents a larger resistance to the aspirating gas flow so that the pressure must be higher than with other angles of attack. An oblique opening in the annulus is advantageous with respect to the actual breakdown of the liquid mist at the upper edge of the solution capillary. If the nozzle of the aspirating gas is tapered toward the tip, an optimum dimension seems to be attained when the half-angle of attack is between 15° and 30°.

Either the aspirating gas orifice can project above the solution capillary, or the reverse can be true (Fig. 5, row 2). The former arrangement is a type of venturi tube with an edge projecting inward which contributes to the reduction in aerosol droplet size (22, 23). If the conical opening around and over the solution capillary is widened (Fig. 5, row 3), always with projecting edges, the effect is further exploited. Alternatively, the solution capillary can be raised above the upper edge of the aspirating gas nozzle, or the heights of the two orifices may be equal. Examples a and c are more suitable for fabrication than case b. Example a has the disadvantage that the interior edge of the aspirating gas nozzle can be rounded off by solid particles in the sample. This lessens the turbulence and efficiency of nebulization. These critical edges may also be damaged when aspirating corrosive solutions. Consequently, type 2c is most frequently used (see also Fig. 13).

The diameter of the aspirating gas orifice can also be varied. Centering the capillary is more critical as the diameter of the aspirating gas orifice is narrowed. With a very small orifice and high gas pressure, the narrow gas jet may not adequately contact all of the liquid, especially the center of the liquid stream.

Finally, the bottom row of Fig. 5 shows that the concentric surface areas can be maintained constant while solution capillaries of varying diameter are provided for passage of the test solution. In emission methods only a limited amount of liquid may be added without lowering the flame temperature significantly, and a solution capillary of narrow diameter would be preferable. Since the flame temperature does not play as decisive a role in atomic absorption, nebulizers with wider bore capillaries can be used profitably. Selection of interior capillary diameter is also affected by the solution being aspirated. For viscous solutions a wide bore capillary would be selected.

All the various possibilities discussed can be used separately and in combination. This provides a variety of possibilities for nebulizer construction.

Materials of construction must be selected to resist corrosion and attack from test solutions. Nebulizers of glass, plastic, brass, stainless steel, platinum, and other noble metals have been described (25–29). Nebulizers fabricated of glass are difficult to reproduce and to maintain tolerances. Different materials are often used for the solution capillary (palladium) and the remainder of the nebulizer (brass or stainless steel). Good care of the nebulizer is necessary. This entails frequent and careful cleaning of the capillary and nozzle. After a series of aspirations the equipment should be rinsed by aspirating distilled water.

C. LESS COMMON TYPES OF PNEUMATIC NEBULIZERS

1. Reversed Concentric Nebulizers

The reversed concentric nebulizer is constructed like the concentric nebulizer; the essential difference is that the aspirating gas flows through the middle capillary rather than the sample solution which enters through the outer concentric annulus. Lundegårdh initiated his work with a nebulizer of this type (30–32). However, the small narrow annulus is more easily clogged than a smooth circular capillary by particles in the liquid, and foreign bodies causing obstructions are more apt to be in the liquid rather than the gas lines.

2. Hydraulic Nebulizer

In the hydraulic nebulizer the liquid is forced under pressure through a small orifice as a jet against a surface barrier where it shatters into droplets. A perpendicular gas stream carries off the droplets thus produced. Liquid compression does not necessarily imply use of a compressed gas so that, strictly, this type of nebulizer could be placed in the nonpneumatic category.

3. Gravity Nebulizer

By contrast with nebulizers that draw up the sample by negative pressure, the gravity nebulizer introduces the sample from a funnel which connects with the capillary of the regular nebulizer. The negative pressure at the nebulization site and the hydrostatic pressure due to the weight of the liquid reinforce each other. Memory effects in changing from one sample to the next are greater because the funnel has a greater surface area than a small suction capillary.

4. Miscellaneous Types

Solids can be aspirated from a trough pulled along underneath a wide-bore capillary of a pneumatic nebulizer, and the powder-air mixture injected directly into the flame (30, 32, 33). In a few instances metal powder, along with oxygen, can be made to burn as a flame when the sample is combustible (*34, 35*). Solids can be pulverized to a fine powder, suspended in a viscous liquid such as glycerin–2-propanol, and the liquid suspension aspirated into the flame using a wide-bore capillary (*36–38*).

D. NONPNEUMATIC NEBULIZERS

Nebulizers with an outside energy source avoid some of the problems that beset a pneumatic nebulizer. However, at the present time only ultrasonic radiation offers a feasible alternative method of nebulization.

1. *Ultrasonic Nebulization*

In Fig. 6 are shown two systems which have been used to produce aerosols (*39*). Power transmission from the transducer to the sample is

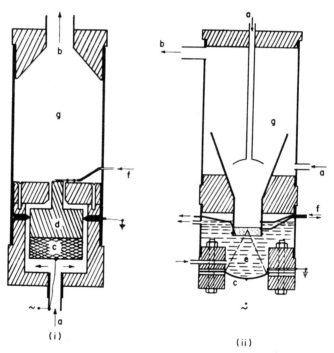

Fig. 6. Ultrasonic nebulizers: (i) velocity transformer and (ii) focusing bowl. a—air, b—aerosol, c—ceramic crystal transducer, d—velocity transducer, e—coupling medium (water), f—sample solution, and g—spray chamber. From (*39*), p. 1354, courtesy of *Applied Optics*.

provided either by a velocity transformer or by a coupling liquid with a focusing device. Sample flow to the ultrasonic nebulizer can be controlled by varying the length of the capillary tubing feeding the nebulizer or by a peristaltic pump.

It has been postulated that standing waves are formed on the surface of the liquid and lead to rupture of sections of the surface to form droplets. Approximate numerical drop size distributions of the aerosols produced are 1.8 μm at 3 MHz, 17 μm at 115 kHz, and 23 μm at 70 kHz. Thus, to produce droplets comparable in diameter with pneumatic nebulizers, ultrasonic frequencies above 500 kHz must be used. Other factors such as cavitation, viscosity, vapor pressure, and ultrasonic power may play a part in the nebulization process, but more likely these parameters affect the nebulization rate rather than the droplet size. An air stream carries the aerosol from the spray chamber into the burner. Contamination from one solution to the next is a problem as unvaporized droplets remain suspended in the vibrational modes of the quartz where they can combine with the next sample (40–45).

2. Electrostatic Nebulizer

In the electrostatic nebulizer the liquid is allowed to rise in a tube through capillary action. A voltage of 5–15 kV is applied between the metal capillary and a counter electrode. The very strong electrical field at the liquid surface produces electric forces which nebulize the liquid. A supplementary air stream carries away the aerosol (46). Sample consumption is small. Since the walls of the spray chamber soon acquire a charge, the electrostatic repulsion prevents later drops from condensing on the walls. However, the rate of nebulization depends on the dipole moment of the molecules of the liquid. Distilled water behaves badly and quite differently from salt solutions, and these even differ depending on their concentration and composition. Many flames become unstable as soon as large amounts of charge carriers are introduced.

3. Chemical Nebulizers

In the chemical nebulizers the test solution is placed in a glass vessel and mixed with chemicals which evolve a gaseous product. The escaping gas bubbles entrain the analyte and burst to generate a mist which is carried into the flame by an air stream (47). This arrangement is easy to improvise but quantitative work is not possible.

4. *Electrolytic Nebulizer*

In electrolytic nebulizers a low dc voltage is impressed between two electrodes dipping into the test solution. Current flows liberating gas which entrains some of the solution. After bursting the mist is carried to the flame by an air stream (*3, 48*).

E. Direct Vaporization into the Flame

Vaporization directly into the flame usually involves methods which are similar to Kirchhoff and Bunsen's early experiments. The powdered or liquid sample is taken onto a platinum loop, dried if liquid, and the wire inserted into the flame. This method finds use for ultramicro determinations in conjunction with integrating circuitry (see Chapter 11 of this volume).

Powdered material can be introduced by the Ramage or "paper spill" method (*49*). A long strip of filter paper is folded to form a channel. The sample, 50–100 mg of powdered material, is spread uniformly along the channel and the folded paper is slowly and uniformly moved into the flame by means of a mechanical device. Of course, the paper can also be impregnated with a known volume (0.05–0.1 ml) of test solution, dried, rolled into a spill, and then moved into the flame.

In both of the foregoing methods, vaporization does not occur uniformly, so that one reads either a maximal deflection or else integrates the signal over the total burning period (*11, 50, 51*). Quasiequilibrium can be attained when a platinum wire is pulled through the test solution, then an air-bath or low temperature oven to dry the sample, and finally through the flame. Reproducibility is poorer than with a nebulizer, and continuous light from the glowing wire can be troublesome. In a slightly different arrangement a platinum gauze disk, on an axis inclined at 45°, is rotated slowly through a dish containing the test solution and tangentially through the flame (*3*).

In the solid-mix method, the flame is replaced by a charge of solid propellant mixed with a known amount of solid sample. The mixture is ignited electrically beneath the optical beam of the spectrometer (*52*). Preliminary sample decomposition, ashing operations, and dissolution are avoided. Sample reproducibility exceeds any other powder method.

III. The Transfer Path

A. The Spray Chamber

The aerosol produced by an indirect nebulizer expands into a spray or expansion chamber where it is mixed with the combustible gas before going to the burner. Within the spray chamber the aerosol droplets desolvate partially and the coarser droplets condense on the walls and are eliminated. What reaches the burner is a "dry" mist homogeneously mixed with the support and combustible gases. There should be a good transfer efficiency with minimal loss of test solution. However, with indirect nebulizers the usual transfer efficiency is 1–3%, occasionally up to 10% (6, 46, 53, 54). Extent of desolvation is limited by the ultimate capacity of the gas phase for holding evaporated solvent, i.e., the saturated vapor pressure at the chamber temperature. The normal capacity for solvent vapor is impaired when the chamber becomes cooled by the desolvation process within the chamber.

A moderate back pressure arises in normal spray chambers from the flow resistance of the aerosol-air mixture in transit from the chamber to the burner and through the exit of the burner head to the flame. This back pressure usually ranges from 10–20 mm of Hg. Pressure variations within the chamber must be smoothed so as not to be converted into signal noise. Since the larger drops are condensed on the walls of the chamber, the liquid film thus accumulated must be eliminated through a drain provided at one bottom end of the chamber, while at the same time maintaining a uniform internal pressure. A typical arrangement is shown in Fig. 7. A sealing liquid is placed in the siphon at the beginning of a series of measurements.

Baffles or spoilers are often built into the spray chamber to improve exchange between the aerosol and the moisture-covered walls. Spoilers also cause the gas and droplets to whirl more intensely which improves aerosol homogeneity.

B. Methods for Improving Transfer Efficiency

Systematic investigations on the construction and dimensions of spray chambers have been made (8, 55–57). Larger spray chambers, preferably

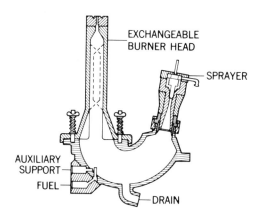

Fig. 7. Indirect nebulizer and spray chamber, Autolam burner (courtesy of Beckman
Instruments, Inc.).

spherical, compel the aerosol droplets to traverse a longer path before
reaching the walls and smooth out pressure variations. This offers more
opportunity for the larger droplets to reach the flame due to a longer period
for solvent evaporation. The larger chamber volume can also assimulate
more solvent vapor. On the other hand, this gain in transfer efficiency is
obtained at the expense of a longer time interval required to attain
equilibrium upon aspirating a fresh solution, and thus a larger supply of
test solution is consumed.

More extensive vaporization of the solvent can be obtained by heating
the walls or interior of the chamber. Using more easily vaporized organic
solvents is also helpful although not always feasible. Three effects play a
role in the heating process. The droplets arriving at the hot walls undergo
the Leiden frost phenomenon—they will dance on the hot surface,
disperse into smaller droplets, and be more completely desolvated. A
heated wall radiates infrared radiation inward, so that the drops in the
interior of the chamber also absorb heat and thus desolvate faster. Finally,
the higher temperature enables more solvent to be vaporized before
saturation is attained. Heating can be done directly by an electric heating
coil, infrared radiation, or high-frequency energy. Although the transfer
efficiency is improved, the disadvantages are serious. Since larger drops
are forcefully vaporized, the solute concentration conveyed to the flame
is not as uniform as with cold chambers. Not only does more analyte reach
the flame, but also more unwanted vaporized solvent. This extra solvent

will lower the flame temperature which may be detrimental. Solids may deposit on the walls of the chamber and memory effects are more troublesome. As the chamber is cooled during the nebulization step, a temperature cycle may arise and a time interval is required for the aerosol to regain its ambient temperature. Because nebulization is not usually carried out in uniform time intervals, rather a period of nebulization and then a pause, thermal creep effects, if present, will cause a waxing and waning of chamber temperature leading to poor reproducibility in the final signal. Condensed-phase interferences are more severe because larger size solute particles are introduced into the flame. These undesirable effects can override any gain in transfer efficiency.

Should it not be desirable or possible to heat the chamber, the aspirating gas can be conveyed over heated metal coils before entering the nebulizer. With preheated gas, subsequent vaporization of solvent will be more complete. However, in addition to the disadvantages already enumerated, the different materials of construction used in the nebulizer will have different coefficients of thermal expansion. Also the nebulizer becomes heated and the critical nozzle adjustments change in various ways depending on the temperature. Preheated compressed air heats the intake solution capillary so that the test solution becomes heated, and this could lead to release of gas bubbles in the test solution and thus erratic aspiration. Certain test solutions, particularly those containing protein, will not tolerate heating directly.

Hot chamber methods combined with liquid condensation overcome some of the disadvantages of hot chamber methods. A cooled impact surface is inserted in the transfer path to remove a portion of the liquid vapor (Fig. 8). Essentially only the liquid constituents are removed while the solid components pass into the flame. Sensitivity is improved and noise fluctuations are reduced; the average improvement is an order of magnitude (58). However, the detection limit is poorer.

A vortex chamber increases transfer efficiency without the use of heat or especially large chambers. A high velocity stream of compressed air is blown into the spray chamber through supplementary nozzles arranged tangentially to the chamber walls. Larger drops which would normally gravitate to the walls are caught up by the air vortex and held in spiral orbits for a rather long time. Thus the droplets are more extensively desolvated and exhibit less tendency to condense on the walls. Despite "dilution" of the original aerosol by the compressed air which contains no aerosol, a gain in atomic concentration entering the flame of 2–3 fold is realized (8).

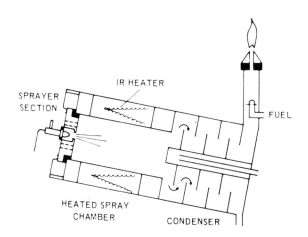

Fig. 8. Diagram of a burner system with radiation-heated spray chamber and condenser (courtesy of Beckman Instruments, Inc.).

C. INJECTION OF FUEL

The inlet for the combustion gas is usually between the spray chamber and the burner, although occasionally it is introduced directly into the spray chamber. The aerosol–support gas mixture and combustion gas must be thoroughly mixed to provide a homogeneous mixture entering the burner so that flame temperature and excitation will remain steady. Also the eddying gas-aerosol mixture must be rendered laminar to yield a quiet flame. Most gas mixing nozzles satisfy the first requirement. The second requirement is fulfilled by a long conduit leading to the burner in which baffles or spoilers are often inserted.

The fuel gas nozzle must not be inserted directly into the aerosol-air stream or salt encrustation would occur. Either the gas nozzle is recessed in the side of the mixing vessel or the gas is introduced tangentially into the mixing chamber; the swirling turbulence promotes mixing, and the aerosol is prevented from reaching the fuel nozzle. Interchangeability of nozzles is recommended so that various combustion gases can be used.

Optimum fuel flows differ due to different heating values. For example, propane nozzles have larger diameter openings than do acetylene nozzles. The quantity of fuel gas to be mixed, the inner diameter and length of the mixing nozzle, and the positive pressure in front of the nozzle selected are substantially dependent on the type of fuel gas and on the design and dimensions of the burner. The advantages and drawbacks of combustion gases have been discussed in Chapter 6 of Volume 1. Here the technical aspects of the equipment will be discussed.

Compressed gas cylinders can be difficult to obtain in remote locations. It may then be advisable to use a "gasoline" gas for the spectrometer operation since fuel depots for gasoline are more prevalent. By using an apparatus similar to an automobile carburetor nozzle, a combustible gas mixture can be produced for a flame spectrometer (59).

IV. Burners

Burners can be divided into two major categories based on the combustion flame and the manner by which the combustible and support gases are mixed. When the fuel gas is mixed with the oxidizing gas before reaching the combustion zone, a premixed flame is produced; to provide this type of flame one uses a premixed or laminar flow burner. When the two gases are mixed in the flame itself, a diffusion flame results. In the latter arrangement the flame turbulence gives rise to the burner name—turbulent burner—also denoted a sprayer-burner because the tip of the nebulizer is located at the point where the gases are mixed.

Within a certain stability range every burner produces a moderately constant flame (60, 61). This stability region is a function of burner construction and its cooling, the kind of gases used and their mode of mixing, and the quantity of gases supplied. Flame stability suffers from too large or too small a volume of combustible and support gas mixture. When the gas supply is excessive, the flame front is lifted off the burner head by the excessively fast gas stream. When the gas supply is too small, the flame burns close to the burner tip, the burner becomes overheated, and vaporization of the burner head may occur. An overheated burner head can often ignite a premixed gas mixture within the burner chamber resulting in flashback of the flame which can destroy the unit.

A. LAMINAR FLOW BURNER

Burners providing premixed flames are based on the design of the well-known Bunsen burner. The basic elements of this type of burner are shown in Fig. 9. Fuel enters via a nozzle just below a Venturi throat whose function is to entrain air or oxygen with the fuel gas and to impart a velocity to the gases which aids in mixing. This is followed by a cylindrical burner body of sufficient length to ensure thorough mixing of the gases, and finally a burner orifice at which extremity the premixed flame rests. A typical burner head consists of a block or tube of chemically resistant material containing either an array of holes or a slot at which the flame burns with little audible noise. An array of holes serves well for low-

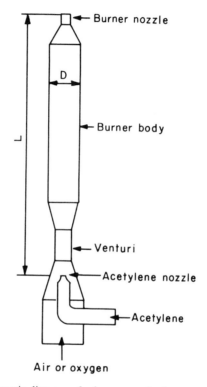

Fig. 9. Schematic diagram of a burner producing a premixed flame.

burning velocity, low temperature flames, such as air–propane, but the rapid encrustation of very small holes required for high-burning velocity, high temperature flames has lead to almost universal use of slot burners for high temperature premixed flames.

1. *Meker Burner*

Designed on the same principle as the Bunsen burner, the Meker burner is an open tube but with a perforated metal head plate affixed to the exit end of the stainless steel burner tube. The greater the depth of openings in the grid, the more nearly laminar will be the flow, and the more stable the flame. The primary or inner combustion zone is distributed as a number of separate small cones, one centered above each opening in the metal grid. The secondary combustion zone, or outer flame mantle, is a common homogeneous cone over the entire burner head. The number of openings in the grid, their size, the heat conductivity of the burner material, and the radiative and conductance cooling of the burner material are all interwoven with the burning velocity and ignition temperature of the gas mixture. One design consists of a square pattern of 17 holes, 0.75 mm in diameter, drilled into the burner head plate.

The flame is prevented from flashing back at the walls of the burner port, where the stream velocity is zero, by the quenching effect of the walls. Since the quenching effect of the walls extends some distance out from the walls, there is a limiting size of burner port or grid opening below which the flame of a given gas mixture will not flash back. The production of a stationary flame requires that the stream velocity of the fuel-oxidant mixture through the burner port be at least equal to the burning velocity. If this condition is not met, the flame is liable to flash back down the burner port and cause an explosion in the spray chamber. Both to ensure a margin of safety and to achieve a reasonably stiff flame, the burner is designed so that the stream velocity of the unburned gases is several times the burning velocity (*3, 60, 61*).

2. Slot Burner

A typical slot burner is shown in Fig. 10. This type of burner produces a laminar premixed flame over a 10-cm slot, single or triple, 0.4–0.6 mm in width, in a burner head fabricated from aluminum or stainless steel. Selection of the proper slot width (and length), considerations involving the heat conductivity of the burner material, and depth of the slot openings are the same as for the Meker burner. However, the slot width is usually narrower than the corresponding hole diameter in a Meker burner, even when the gas mixture and burner properties are the same. Cooling in the slot is not as satisfactory as in a hole due to the unfavorable ratio of surface area to volume. A three-slot (Boling) burner provides a wider encirclement of atomic vapor that more fully encloses the optical beam. The outside sheath of flame is not outside the optical path and yet thermally insulates the inner sheath while also protecting it from entrained air. Fuel adjustment is less critical.

Slot burners are preferred for atomic absorption work because a long absorption path through a quiet homogeneous flame is important. To obtain high solute concentrations for the longer slot length, very thin flames are used. Interchangeable spacers can be inserted to adapt the slot width to a particular gas mixture. A peaked, slanted, moveable slot allows the opening to be varied according to the fuel-oxidant selected. Usually a

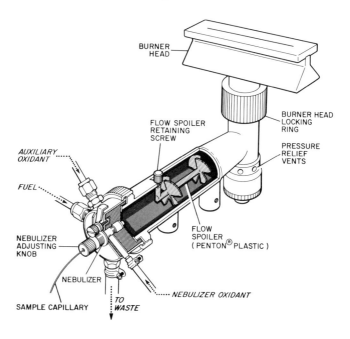

Fig. 10. A slot burner and expansion chamber (courtesy of Perkin–Elmer Corporation).

10-cm slot length is used for air–acetylene flames, but a 5-cm slot is needed for nitrous oxide–acetylene flames. A baffle inserted beneath the slot provides a more uniform distribution of the aerosol along the length of the slot. In a less common design, the aerosol and air exit from several large central apertures and the combustible gas exists from a rectangular line of separate smaller openings surrounding the central apertures (62). This design offers complete safety from flashback. However, the burner surface is subject to encrustation since the aerosol desolvates more slowly. Also the mixing and combustion of the fuel gases is less efficient.

3. *Burners for Separated Flames*

In separated flames the inner combustion zone is deliberately separated from the surrounding diffusion flame, as shown in Fig. 11. Using a wide silica tube fitted over the Meker burner, the flame separator forms an

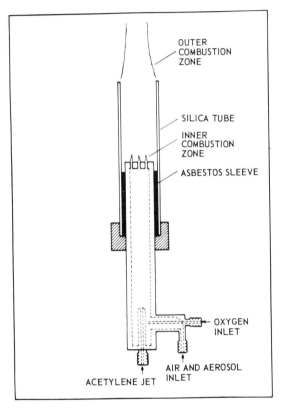

OUTER
COMBUSTION
ZONE

SILICA TUBE

INNER
COMBUSTION
ZONE

ASBESTOS SLEEVE

OXYGEN
INLET

AIR AND AEROSOL
INLET

ACETYLENE JET

Fig. 11. Burner arrangement for a separated air–acetylene flame employing a silica separator tube.

extension above the inner burner port (*4, 62, 62a*). The primary combustion then occurs at the inner burner port, while the pale blue secondary diffusion flame is maintained at the top of the outer silica tube. By this method the interconal zones of flames are extended in length. Stable separated flames are obtained over a fairly wide range of fuel-air mixture strengths. Owing to the extended width and height of the reducing interconal zone, the absorbance (and emission) signals obtained for refractory elements

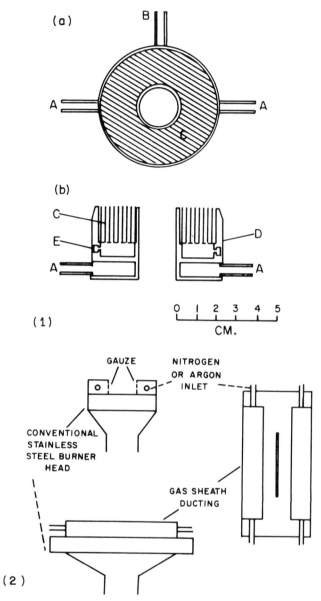

Fig. 12. (1) Diagram of shielded flame emission burner: (a) top view; (b) side view; (A) water pipes; (B) nitrogen–argon inlet; (C) laminar flow matrix; (D) support for matrix; (E) nitrogen–argon distribution channel. (Reprinted from (*62*) by courtesy of *Laboratory Practice*.) (2) Diagram of shielded atomic absorption burner for nitrous oxide–acetylene flame. (Reprinted from (*65*) by courtesy of Applied Optics.)

introduced into the flame is less dependent on the position and focusing of the flame relative to the optical axis of the source and monochromator.

An alternative method of separation of premixed air–hydrocarbon flames consists of sheathing the flame with an inert gas (nitrogen or argon) to lift off or to separate the secondary diffusion zone. In a typical design (Fig. 12) the shielding gas reservoir is constructed of brass or aluminum. Nitrogen or argon is passed into a distribution manifold in the outer wall which has several equally spaced outlets around its inner face. The gas then passes vertically through many hundreds of narrow channels of triangular cross section and emerges from this matrix with a laminar flow that persists for a distance of 4–5 cm when the flow rate is adjusted to 10–15 liters/min (63–65). A modification for the nitrous oxide–acetylene flame has also been described (62). This arrangement has several advantages over the mechanical method of separation with a silica tube. A wider range of fuel-air mixture strengths may be used, and less difficulty is encountered owing to carbon deposition when organic solvents are aspirated into the flame. Slot burners have been sheathed by placing gauze-filled ducting parallel to the burner slot. The inert gas emerges through the gauze alongside the flame and just above the primary zone, separating the secondary zone from it.

With sheathed flames the interconal zones of flames are extended in length and exhibit very low radiative background. The interconal zone contains the hottest part of the flame and can be viewed without interference from radiation produced in a secondary diffusion that would normally surround it.

B. TURBULENT (SPRAYER) BURNER

The characteristic feature of a turbulent flow burner is that the fuel gas and the support gas are not mixed until the point at which they enter the flame; nebulization also occurs at this point. Thus, the unit is a combination nebulizer-burner. It is also referred to as a direct injection sprayer-burner since all the liquid aspirated enters the flame and is converted to a spray at the point of entry. This type of burner is inherently safer since the fuel and support gases do not mix until the actual point of entry into the flame; thus flashback is impossible. A drawback is the loud audible noise which the flame emits when burning and which arises from the turbulence created as the gases mix and burn in the same local area.

1. *Nebulizer-Burner*

Figure 13 shows a typical nebulizer-burner combination. It produces a diffusion flame supplied by direct nebulization of the sample into the flame. If the flow velocities of the support and combustible gases are very large, so that the dimensionless Reynold's number, Re, exceeds 3200, where

$$Re = vd/\eta \tag{1}$$

and v is the flow velocity, d is the diameter of the tube or nozzle, and η is the kinematic viscosity, i.e., viscosity/density, we are dealing with

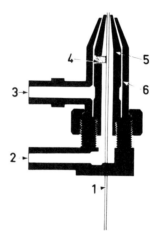

Fig. 13. Nebulizer-burner unit: 1, solution capillary; 2, aspirating gas inlet; 3, fuel gas inlet; 4, centering screw; 5, gas inlet and jacket; 6, jacket (courtesy of Beckman Instruments, Inc.).

turbulent flow where the gas streams are no longer parallel but whirl intensely against one another. This turbulence creates a very noisy flame.

A diffusion flame sits directly on the outer edge of the outermost concentric gas nozzle. This small, symmetrical outer diffusion flame serves as a pilot light for the main combustion mixture. It is safe to use for flames of high-burning velocity such as oxygen and acetylene (or hydrogen). A narrow oxygen opening is desirable. Centering the solution capillary in the oxygen nozzle is quite critical. Use of centering screws runs the risk of having oxygen escape into the gas volume through leaks along the screws thus producing a combustible gas mixture prematurely. Another design uses spacers which are slipped on or soldered on the capillary.

Atomization in flames from nebulizer-burner units is frequently in-efficient, and interferences are more severe because relatively large droplets are introduced directly into the flame. Desolvation may not be sufficiently rapid in the short passage time through the preheating zone beneath the interconal gases.

2. Reversed Flame

In nebulizer-burners one can reverse the combustible and support gas so that nebulization is carried out with the combustible gas (9, 66–68). In this mode the flame lifts off the burner tip and thus the burner is not so easily heated, an advantage when working with solutions which are subject to solute coagulation. The reversed flame is not as stable and drop size distribution is poorer.

3. Sheathed Burners

The flame can be surrounded by a protective sheath of either oxygen or compressed air (Fig. 14) (9). Care must be taken that the flame is not rendered aslant by unilaterial addition of the protective gas, such as might occur by tangential injection of the gas at the lower end of a tube open

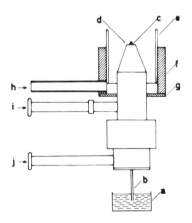

Fig. 14. Sheathed nebulizer-burner unit: (a) test solution; (b) sample capillary; (c) oxygen nozzle; (d) fuel nozzle; (e) copper sheet cylinder; (f) brass cylinder; (g) brass washer; (h) oxygen (or air); (i) fuel; (j) oxygen (or air) (courtesy of Beckman Instruments Inc.).

at the top. Advantages of sheathed flames are greater flame stability, less entrainment of room particles into the flame, and higher flame temperatures through entrainment of oxygen instead of air in the outer combustion zone.

4. *Burners for Entrained Air–Hydrogen Flames*

Hydrogen is fed through a portal supplying 24 small holes arranged in a circle on the ring head. Air is delivered from a separate supply tube which entered into an outer mantle of the nebulizer and, ultimately, the outer space of its concentric orifice. Sample solutions are aspirated pneumatically through the inner tube of the nebulizer by means of nitrogen, argon, or helium to be delivered to its inner orifice or nebulized into the air–hydrogen flame. The flow rate of sample solution is held in the range 2.0 ± 0.5 ml/min at 15 psi (2.5 liters/min) of nitrogen or argon.

5. *Less Common Types*

Improvement in the performance of the sprayer-burner is achieved by premixing the fuel and oxidant in the two outer concentric tubes, thereby making the flame more laminar and producing well-defined zones (*69*). As illustrated in Fig. 15, the gases are introduced into each port of a Jarrell-Ash HETCO burner and are throttled through small, premixing, dual needle valve assemblies, one for each port. Assembly and operating directions are given in the reference cited.

All three components need not be fed to the burner cap unmixed. The test solution can be sprayed directly into the center of a flame with support gas while a fuel-rich gas mixture exits around it from a separate nozzle or slot (Fig. 16). The disadvantage of this arrangement is the necessity for water cooling and the fact that an explosive mixture must be prepared in a mixing chamber ahead of the burner (*3, 5, 70*).

In principle, it is possible to spray the sample directly into the flame with a prepared mixture of combustible and support gas (*6*). Again the preliminary preparation of an explosive mixture is hazardous. In these designs there are fewer possibilities for using long, thin holes or slots as heat conductors to minimize flashback.

C. TRANSITION AND SPECIAL BURNERS

A large part of the problem of the turbulent flame arises from disruption of the suitable chemical environment found in the well-defined

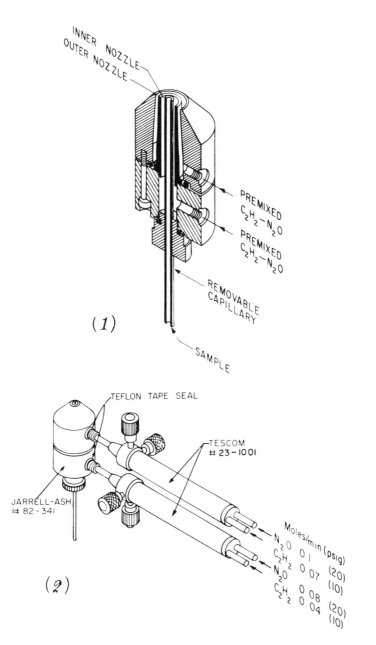

Fig. 15. (1) Cross-sectional view. (2) Premixed nebulizer-burner unit (reprinted from (69) by courtesy of *Applied Optics*).

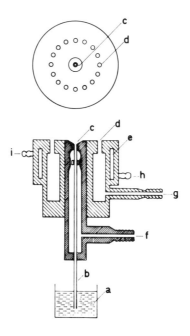

Fig. 16. Weichselbaum-Varney direct nebulizer-burner: (a) test solution; (b) solution capillary; (c) nebulizer opening; (d) burner port; (e) water jacket; (f) aspirating gas inlet; (g) premix gas inlet; (h) cooling water inlet; (i) cooling water outlet. From (*70*).

zones of the premixed flame. Considerable improvement in the performance of a typical sprayer-burner is provided by adding a small premixing chamber between the surface where the gases emerge from the concentric tubes and the flame (*23*). The more refined version possesses a stainless steel insert in a Teflon barrel. Flooding, a problem in the earlier graphite version, is reduced since excess nebulized solution readily drains from the tip into the reservoir around the top of the barrel. Contamination, or memory effects, are minimized. However, the stainless steel insert must be protected from excessive heat by continuous nebulization to provide adequate cooling of the burner tip and to maintain the solvent level in the reservoir. By suitable selection of the insert diameter and method of sprayer-burner operation, a laminar premixed oxygen–acetylene flame can be maintained over the tip of tube (Fig. 17). This type of flame requires that the aspirated solution contain approximately 90% organic solvent (usually ethanol).

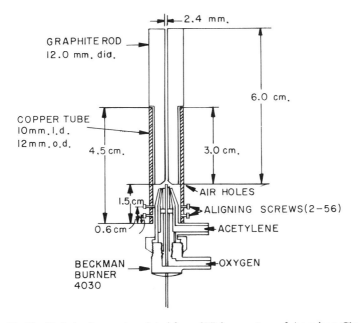

Fig. 17. The Kniseley burner (reprinted from (*23*) by courtesy of American Chemical Society).

V. Operational Techniques

The nebulizer must be adapted to the burner so that optimal operation parameters for each conform within the stability limits of the flame. These parameters include the optimum support gas pressure and flow rate, including also the aspiration rate of the test solution. Practically, this means that burners and nebulizers cannot be interchanged indiscriminately.

A. Advantages and Disadvantages of Laminar and Turbulent Burners

Several advantages accrue from the separation of nebulizer from the burner. When one component fails, repair or replacement difficulty and expense is less. Nebulization can be controlled independently and a nearly uniform drop-size distribution is fed to the flame after passage through the spray chamber. Less solvent reaches the flame. The flame is more homogeneous, the height dependence for observation is not so critical, at least within the secondary combustion zone. In fact, the entire desolvation, vaporization, atomization, and excitation processes are smoother since the larger drops are rejected before reaching the flame and thus do not create local inhomogeneities. Scattering of background radiation and radiation from hot particles is minimized; this is significant in atomic fluorescence.

Some disadvantages exist. Cooling from evaporation of solvent within the spray chamber is a factor which can affect the transfer efficiency. There is a rather long time lapse of several seconds between the initiation of nebulization and the onset of an equilibrium state within the chamber and in the flame gases.

An integral sprayer-burner is compact and obviates variables affecting transfer efficiency. Flashback is unlikely and thus flames with high burning velocity and organic solvents may be used. All the aspirated liquid enters the flame; however, the vaporization or desolvation efficiency is poor (5, 6, 54, 71). The larger droplets fail to desolvate and the solid within to vaporize and be converted into atomic vapor in the short residence time within the flame. Disturbances, such as condensed phase interference, are more severe. Flame temperature is seriously affected by the greater heat loss to desolvation within the flame gases. Encrustation of salts or carbon buildup from fuel-rich hydrocarbon flames are sometimes troublesome factors.

From the beginning of atomic absorption work, the slot burner, providing a rather long absorbing path, was employed. However, recognition of the fact that line emission intensities are influenced in the same way by flame length is more recent (72). Of course, increasing flame emission intensities in this way is of little use in the presence of strong flame background intensity. Often, though, the greater line intensities obtained with slot burners permit the use of narrower slit widths, with predictable and worthwhile gains in the line-to-background ratio in certain practical situations.

B. MEANS FOR LENGTHENING THE ABSORPTION (AND EMISSION) PATH

1. Multiple Traversals of Light Source

A longer absorption path can be achieved by using a system of mirrors which reflect the source radiation back-and-forth 3 or 5 times through the flame (Fig. 18) (*8, 73, 74*). However, the absorption does not increase in proportion to the number of passes through the flame because the source radiation cannot be passed at the same observation height on each passage. Thus, some of the passages will not intersect the flame at the optimal observation height (see Chapter 8 of Volume 1). In increasing the absorption path, the noise component of the signal also increases. Furthermore,

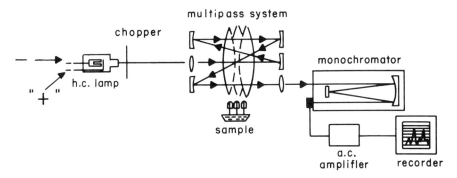

Fig. 18. Multiple traversal of the flame by light source (courtesy of Jarrell-Ash).

in the far ultraviolet portion of the spectrum, the number of passes is limited because of the poor reflectivity of mirror surfaces and the lower transmission of these wavelengths through the flame gases.

2. *Long-Path Burner*

To provide a longer optical path through the flame of a turbulent sprayer-burner, the flame can be introduced at the base of a "tee" pipe which forces the flame to spread out along the optical axis. Alternatively, the sprayer-burner is inclined at a 45° angle and the oblique flame pointed into the end of a horizontal quartz or ceramic tube (1 cm inner diameter) aligned on the optical axis (*75*). To define the absorption path, an intense air stream is blown vertically at the farther open end of the tube or pipe to deflect the flame gases upward (Fig. 19). The Fuwa absorption tube, as this arrangement is denoted, improves detection limits for turbulent burners for all metals whose oxides are easily dissociated. The success of the arrangement depends on the mean lifetime of free atoms and this in turn on the temperature and composition of the flame gases. Heating the absorption tube offers increased sensitivity for certain elements and reduced interferences.

Critical in this method is the atomic recombination rate (see Chapters 4, 8, and 11 of Volume 1). The lifetime of free atoms may be limited by two basically different processes: (1) Condensation of the metal if the partial pressure of the vapor at the given temperature is exceeded, and (2) reactions of the free atoms with the different gas components. Vapor pressures may be exceeded only for relatively involatile elements such as platinum or rhodium. Among reactions, that with oxygen is generally most important. For those elements which tend to have their atoms recombine quickly in the flame with atomic oxygen or other entities, the

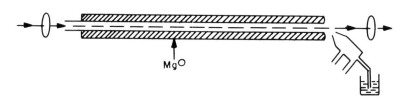

Fig. 19. Fuwa absorption tube (reprinted from (75) by courtesy of American Chemical Society).

gain in absorbance will be slight. The fuel-oxidant ratio involves also the air entrained in the absorption tube from surrounding atmosphere. For a given tube diameter, the amount of entrained air depends on the angle between the burner and the optical axis. The larger this angle, the less air enters the tube and the lower is the necessary fuel flow. The actual state resembles a split flame along a horizontal axis.

Problems arise in maintenance and cooling of the absorption tube. Memory effects are troublesome, the difficulty increasing with decreasing vapor pressure of the test element (76–78). Undoubtedly reflections occur from the inner surfaces of the tube; these can be affected by salt deposition or general corrosive effect of flame gases.

3. Burners in Series

Several sprayer-burners can be arranged in series along the optical axis so that perhaps three flames are burning behind one another. The absorption path can be varied within broad limits. The difficulty lies in adjusting two or three nebulizer-burners to behave in an identical manner.

All the methods used to improve atomic absorption response by extending the absorption path are just as effective in improving emission intensities. The improved sensitivity allows narrow spectrometer slits to be used with a corresponding improvement in the line-to-background ratio (*72*).

C. INTERMITTENT SAMPLE ADDITION

Normally aspiration is continuous while each solution is being sprayed. However, there are situations where it is desired to aspirate periodically samples with pauses in between. Aspiration and pause will follow one another in rapid succession. The flame background will be present during the pauses as well as during the aspiration periods. A steady dc signal is therefore emitted by the background while the emission features from the test solution produce an ac signal because of the intermittent aspiration. Consequently, insertion of an ac amplifier after the detector will result in suppression of the flame background signal (*24, 51, 79*). In atomic absorption and atomic fluorescence only lines actually absorbed by the flame plus analyte aerosol produce an ac signal (*80, 81*). Nonabsorbing lines are not included, and this provides improved optical resolving power.

In a slightly different operating mode, two or more solutions are aspirated periodically, alternating each in rapid succession. All signals from a particular solution, after passage through a signal splitter, go into a single measuring channel. For example, a typical set of solutions might be the test solution (A), a blank solution (B), and a standard solution (S). The integrated values of (S − B) and (A − B) can be produced electronically and the quotients of these differences can be ascertained by analog or digital computers to give a rapid calibration method (*24, 51, 79–81*). Unwanted flame lines are suppressed better than by intermittent aspiration. Variations in the background radiation from the flame or source in atomic absorption can be avoided as long as the noise fluctuations are slower than the periodically recurring process of aspiration. Practical realization of these promising methods is hampered by technical difficulties.

REFERENCES

1. J. A. Dean, *Flame Photometry*, McGraw-Hill, New York, 1960.
2. R. Herrmann and C. T. J. Alkemade, *Chemical Analysis by Flame Photometry*, Wiley-Interscience, New York, 1963.
3. R. Mavrodineanu and H. Boiteux, *Flame Spectroscopy*, Wiley, New York, 1965.
4. R. L. Mitchell, *Commonwealth Bur. Soil. Sci. (Gt. Brit.) Tech. Commun.*, **44**, 1948.
5. B. L. Vallee and R. E. Thiers in *Treatise on Analytical Chemistry* (I. M. Kolthoff and P. J. Elving, eds.), Part I, Vol. 6, Wiley-Interscience, New York, 1965, p. 3463.
6. R. Herrmann, *Z. Aerosol-Forschung u Therapie*, **3**, 1 (1954).

7. R. Herrmann and W. Rick, *Acta Geol. Geograph. Univ. Comenianae Geol.*, No. 6, 507 (1959).
8. R. Herrmann and W. Lang, *IXth Colloquium Spectroscopicum Internationale, Lyon*, G.A.M.S., Paris, 1962, pp. 291–308.
9. P. T. Gilbert, Jr., *Am. Soc. Testing Mater. Spec. Tech. Publ. 269*, 73 (1960).
10. P. T. Gilbert, Jr., *Xth Colloquium Spectroscopicum Internationale* (E. R. Lippincott and M. Margoshes, eds.), Spartan, Washington D.C., 1963, pp. 171–215.
11. J. Isreeli, M. Pelavin, and G. Kessler, *Ann. N.Y. Acad. Sci.*, **87**, 636 (1960).
12. D. P. Muny, *At. Abs. Newsletter*, **3**, 129 (1964).
13. F. J. Feldman, *Am. Laboratory* (August 1969), p. 47.
14. K. M. Cellier, *Appl. Spectry.*, **20**, 26 (1966).
15. P. T. Gilbert, Jr., *Anal. Chem.*, **34**, 210 (1962).
16. J. A. Dean and W. J. Carnes, *Analyst*, **87**, 743 (1962).
17. J. A. Dean and J. E. Adkins, Jr., *Analyst*, **91**, 709 (1966).
18. S. Eckhard and A. Püschel, *Z. Anal. Chem.*, **172**, 334 (1960).
19. K. Fuwa, R. E. Thiers, B. L. Vallee, and M. R. Baker, *Anal. Chem.*, **31**, 2039 (1959).
20. R. Püschel, L. Simon, and R. Herrmann, *Optik*, **21**, 441 (1964).
21. J. D. Winefordner and H. W. Latz, *Anal. Chem.*, **33**, 1727 (1961).
22. D. A. Davies, R. Venn, and J. B. Willis, *J. Sci. Instr.*, **42**, 816 (1965).
23. R. N. Kniseley, A. P. D'Silva, and V. A. Fassel, *Anal. Chem.*, **35**, 910 (1963); **36**, 1287 (1964).
24. R. Herrmann, *Z. Instr.*, **75**, 101 (1967).
25. B. S. Marshall, *J. Sci. Instr.*, **43**, 199 (1966).
26. G. E. Peterson, *At. Abs. Newsletter*, **5**, 142 (1966).
27. V. C. O. Schüler and A. V. Jansen, *J. S. Afr. Inst. Min. Mater.*, **62**, 790 (1962).
28. V. C. O. Schüler, A. V. Jansen, and G. S. James, *J. S. Afr. Inst. Min. Mater.*, **62**, 807 (1962).
29. M. Zeldman, *At. Abs. Newsletter*, **6**, 50 (1967).
30. H. Lundegårdh, *Die Quantitative Spektralanalyse der Elemente*, Fischer Verlag, Jena, Part I, 1929; Part II, 1934.
31. H. Lundegårdh, *Lantbruks.-Hogskol. Ann.*, **3**, 49 (1936).
32. H. Lundegårdh, *Metallwirtschaft*, **17**, 1222 (1938).
33. S. A. Shipitsyn, V. V. Kryushkin, and N. Y. Kuklena, *Zhur. Anal. Khim.*, **21**, 779 (1966).
34. R. Edse, K. N. Rao, W. A. Strauss, and M. E. Michelson, *J. Opt. Soc. Am.*, **53**, 436 (1963).
35. T. Kantor and L. Erdey, *Talanta*, **13**, 1289 (1966).
36. P. T. Gilbert, Jr., *Anal. Chem.*, **34**, 1025 (1962).
37. B. Krause, (Beckman) Report 3/4, 9 (1966).
38. J. L. Mason, *Anal. Chem.*, **35**, 874 (1963).
39. J. Stupar and J. B. Dawson, *Appl. Opt.*, **7**, 1351 (1968).
40. J. A. Dean in *Developments in Applied Spectroscopy* (L. R. Pearson and E. L. Grove, eds.), Vol. 5, Plenum Press, New York, 1966, p. 317.
41. H. Dunken, G. Pforr, W. Mikkeleit, and K. Geller, *Z. Chemie*, **4**, 237 (1964).
42. H. Dunken, G. Pforr, W. Mikkeleit, and K. Geller, *Spectrochim. Acta*, **20**, 1531 (1964).
43. H. C. Hoare and R. A. Mostyn, *Anal. Chem.*, **39**, 1153 (1967).

44. C. D. West, *Anal. Chem.*, **40**, 253 (1968).

45. W. J. Kirsten and G. O. B. Bertilsson, *Anal. Chem.*, **38**, 648 (1966).

46. H. Straubel, *Mikrochim. Acta* (1955) 329.

47. T. Török, *Z. Anal. Chem.*, **116**, 29 (1939); **119**, 120 (1940).

48. E. Beckmann, *Z. Elektrochem.*, **5**, 327 (1899).

49. H. Ramage, J. H. Sheldon, and W. Sheldon, *Proc. Roy. Soc. (London)*, **B113**, 308 (1933).

50. L. R. P. Butler and A. Strasheim, *Spectrochim. Acta*, **21**, 1207 (1965).

51. R. Herrmann, *Z. Anal. Chem.*, **212**, 1 (1965).

52. A. A. Venghiattis, *Spectrochim. Acta*, **23B**, 67 (1967).

53. C. T. J. Alkemade, *A Contribution to the Development and Understanding of Flame Photometry*, Ph.D. Thesis, Utrecht, 1954.

54. J. B. Willis, *Spectrochim. Acta*, **23A**, 811 (1967).

55. R. Herrmann and W. Lang, *Arch. Eisenhüttenw.*, **33**, 643 (1962).

56. H. L. Kahn, *Res. Standard*, **5**, 337 (1965).

57. F. H. Schlaser, *Z. Instrumentenk.*, **73**, 25 (1965).

58. A. Hell, W. F. Ulrich, N. Shifrin, and J. Ramirez-Muñoz, *Appl. Opt.*, **7**, 1317 (1968).

59. N. S. Poluektov, *Techniques in Flame Photometric Analysis*, Consultant's Bureau, New York, 1961, pp. 48–51.

60. B. Lewis and G. von Elbe, *Combustion Flames and Explosions of Gases*, Academic Press, New York, 1951.

61. A. G. Gaydon and H. G. Wolfhard, *Flames, Their Structure, Radiation and Temperature*, 2nd ed., Chapman & Hall, London, 1960.

62. D. N. Hingle, G. F. Kirkbright, M. Sargent, and T. S. West, *Lab. Practice*, **18**, 1069 (1969).

62a. S. S. Brody and J. E. Chaney, *J. Gas Chromatog.*, **4**, 42 (1966).

63. G. F. Kirkbright, A. Semb, and T. S. West, *Talanta*, **14**, 1011 (1967).

64. G. F. Kirkbright, A. Semb, and T. S. West, *Spectry. Letters*, **1**, 7 (1968).

65. G. F. Kirkbright and T. S. West, *Appl. Opt.*, **7**, 1305 (1968).

66. W. L. Crider, *Anal. Chem.*, **37**, 1770 (1965).

67. J. A. Dean in *Developments in Applied Spectroscopy* (E. Davis, ed.), Vol. 4, Plenum Press, New York, 1965, pp. 443–455

68. R. K. Skogerboe, A. T. Heybey, and G. H. Morrison, *Anal. Chem.*, **38**, 1821 (1966).

69. V. G. Mossotti and M. Duggan, *Appl. Opt.*, **7**, 1325 (1968).

70. E. Weichselbaum and P. L. Varney, *Proc. Exp. Biol. Medicine*, **71**, 570 (1949).

71. M. L. Parsons and J. D. Winefordner, *Anal. Chem.*, **38**, 1593 (1966).

72. S. R. Koirtyohann and E. E. Pickett, *Appl. Spectry.*, **23**, 597 (1969).

73. R. Herrmann and W. Lang, *Z. Klin. Chem.*, **1**, 182 (1963).

74. P. B. Zeeman and L. R. P. Butler, *Appl. Spectry.*, **16**, 120 (1962).

75. K. Fuwa and B. L. Vallee, *Anal. Chem.*, **35**, 943 (1963).

76. H. Brandenberger and H. Bader, *At. Abs. Newsletter*, **7**, 53 (1968).

77. J. A. Burrows, C. H. Heerdt and J. B. Willis, *Anal. Chem.*, **37**, 579 (1965).

78. B. V. L'vov, *Spectrochim. Acta*, **17**, 761 (1961).

79. R. Herrmann, *Z. Klin. Chem.*, **3**, 178 (1965).

80. W. Lang, *Z. Anal. Chem.*, **223**, 241 (1966).

81. W. Lang, *Z. Anal. Chem.*, **219**, 321 (1966).

4 Nonflame Absorption Devices in Atomic Absorption Spectrometry

H. Massmann

INSTITUT FÜR SPEKTROCHEMIE UND ANGEWANDTE SPEKTROSKOPIE
DORTMUND, WEST GERMANY

I. Introduction

Flames are normally used as the absorption volume in atomic absorption spectrometry. However, flames are not suitable for some analytical problems, for example, analysis in the vacuum ultraviolet, the detection of extreme traces, and the analysis of very small amounts of samples. In some cases these problems can be solved only when another absorption volume is used in place of the usual flame plasma.

Nonflame volumes used successfully in atomic absorption spectrometry include electrical discharges, especially those in cooled or in hot hollow cathodes. In recent years work which has been done to test electrically generated "flames" includes the plasma torches and stabilized arcs. For trace analysis, high temperature furnaces are becoming important for the detection of very small amounts. Heated graphite tubes become the

TABLE 1
CHARACTERIZATION OF NONFLAME SOURCES

Type	Con-figuration	Sample amount handled	Fill gas pressure	Duration absorption signal	Applications
Electrical discharges	Hot hollow cathode	100 mg	0.5–8 Torr	10–60 sec	Solid samples
	Cooled hollow cathode		0.5–8 Torr	Unrestricted	Isotopes and gases
High temperature furnaces	Graphite tube with additional arc	0.5 mg	1–6 atm	1 sec	Trace and microanalysis
	Graphite tube, no additional arc	0.5 mg	1 atm	5 sec	Trace and microanalysis
Pulsed heating	Heated carbon filament	0.5 mg	1 atm	3 sec	Trace and microanalysis
	Flash lamp		1 atm	1 msec	Microanalysis
	Laser	10–50 μg	1 atm	1 msec	Localized areas

absorption volume. The xenon flash lamp and the solid state laser are also used as radiation sources in the evaporation of analytical samples for atomic absorption. A short characterization of some nonflame sources are listed in Table 1. Solid samples can be vaporized in all absorption cells characterized in this table. In some cases it is more convenient to introduce liquid samples into the absorption cell; however, one must dry the sample before vaporizing the solid constituents. With this technique the sample vapor in the absorption cell is not diluted by the solvent. In all methods tabulated the sample is usually evaporated in an atmosphere of rare gas.

An exception is the evaporation by laser heating although it is possible to use rare gases in this case also. Consequently, all these absorption volumes are suitable for atomic absorption analysis in the vacuum ultraviolet.

II. Absorption Cells

A. HOT HOLLOW CATHODES

In 1959 Russell and Walsh (*1*) suggested that cathodic sputtering might be a means to obtain an atomic vapor from solid samples for use in atomic absorption spectrometry. Gatehouse and Walsh (*2*) were the first to use a hollow cathode tube as an absorption cell. They determined silver in metallic copper in the range of 0.005–0.05 %. The cell had the form of a cylindrical tube, 40 mm in length and 12 mm in internal diameter. The cell fitted in a spring clip and was used directly as the hollow cathode in a stainless steel discharge chamber. The temperature was not controlled and depended on the discharge current. Since the current used was only 60 mA, the cathode remained relatively cool. Later, Walsh (*3, 4*) reported that Sullivan had used discharge currents of several hundred milliamperes, thus obtaining higher temperatures in the hollow cathode, for the determination of phosphorus and silver in copper, and silicon in aluminum and steel. For phosphorus the resonance line at 1774.95 Å was employed, a line in the vacuum ultraviolet where absorption measurements with chemical flames are not possible. In all examples the sample was identical as the hollow-cathode cylinder. This means that unrestricted amounts of sample were analyzed.

Atomic absorption methods with hot hollow cathodes and restricted amounts of samples were first described by Ivanov et al. (*5*). The cathode cylinder was held in the optical path by a thin molybdenum foil. Therefore, the cooling of the cathode due to heat conduction is small in comparison with radiation cooling. The cathode becomes red-hot during the discharge period. The sample, placed in the cathode as a solid or as a solution, is entirely evaporated in the discharge. A similar type of absorption cell is shown in Fig. 1 (*6*). The cathode cylinder is a tube of graphite or some high-melting metal, 40 mm in length, 7 mm in inner diameter, and 1.5 mm wall thickness. The cell was held in the optical path by a 2-mm molybdenum wire in a metallic discharge chamber. Normally the current used is 200–600 mA.

Atomic vapor is produced by boiling the sample in the hot hollow cathode and only to a lesser extent by cathodic sputtering. This can be observed if the discharge current is suddenly interrupted. The absorption signal decreases slowly and at a rate which is dependent on the heat capacity of the cathode. Also, at the beginning of the discharge the absorption signal appears only after the cathode has attained a certain temperature. The emission that is produced by the hollow cathode can

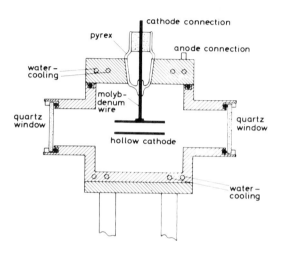

Fig. 1. Hot hollow cathode in a water-cooled discharge chamber.

be strong because of the relatively high discharge currents needed to evaporate the sample in a short period of time. After the initial discharge a strong band emission occurs. This may cause serious spectral interference if the usual measuring technique employing modulation of the primary radiation is not used. These difficulties can be overcome by heating the absorption cathode with a half-wave current and by chopping the light beam between the absorption cell and the spectrometer. In the cycle when the discharge is running and emission occurs, the light path is interrupted by the chopper. In the next half-cycle, when the hollow cathode is dark, the absorption signal can be measured with a tuned amplifier.

Because of the restricted amount of sample, an analytical signal will be time dependent. For example, the determination of silver in metallic lead is shown in Fig. 2. The figure relates the time dependence of the resonance absorption of silver for 5, 20, and 170 $\mu g/g$ during the evaporation of 50-mg samples of lead in a hollow cathode discharge (6). The hollow cathode was a graphite tube, 40 mm in length, held in the optical path by a 2-mm molybdenum wire. The discharge current was 350 mA with an argon atmosphere at 6 Torr. Under these conditions the time required to evaporate the sample is less than 2 min. Depending upon the concentration of silver, not only is the height of the absorption signal changed, but there is a remarkable change in the slope of the curve. Because of these changes the integral of the absorbance signal should be taken for the working curve

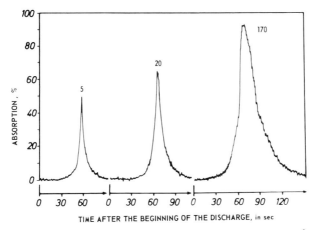

Fig. 2. Time dependence of the absorption of silver, $\mu g/g$, at 3281 Å during the vaporization of 50-mg samples of lead in a hollow-cathode discharge.

rather than the maximum absorbance. If the integral of the absorbance is used, the working curves are linear over a range of several orders of magnitude.

The hot hollow cathode as an absorption cell has a remarkable advantage for the analysis of solid samples. Because of the low pressure the sensitivity in the hollow cathode is less than in the well-known graphite furnaces at atmospheric pressure; however, more than 100 times more sample can be evaporated without interference from unspecific background absorption.

B. Cooled Hollow Cathodes

The absorption cell used by Goleb and Brody (7) is a modified water-cooled Schüler-Gollnow tube. Interchangeable cathode cylinders are fabricated of aluminum and are 6.3 cm in diameter and 2.54 cm in length. The sample is introduced into the cell as a solution and distributed over the entire inner surface of the cathode before being dried under a heat lamp. The hollow-cathode discharge in the absorption tube operates with neon at 2 Torr as the carrier gas. Detection limits for Mg, Ca, Be, and Si are approximately 1 μg. The absorption signal is increased with higher current in the absorption tube. However, faster sputtering at higher currents causes fluctuations in the absorption signal. For the determination of sodium 100 mA was recommended; 27 μg of sodium was determined with a relative standard deviation of 8 %. At the present time only a few studies have been reported using cooled hollow cathodes as absorption cells except for isotopic analysis.

The isotopic compositions of H, He, Li, B, Hg, and U have been determined by atomic absorption spectrometry. Isotopic absorption analysis of mercury has been made in a discharge-free absorption cell (8), helium and hydrogen in discharge lamps (9, 10), and the isotopes of lithium and uranium in hollow-cathode tubes. The absorption cells used were different.

If cooled hollow cathodes are used as the primary source as well as the absorption cell, the emission and absorption lines are very narrow. Consequently, in some cases the lines of different isotopes of an element may be separable, more or less, with a spectrometer of high resolving power, and the concentration of a single isotope determined. Normally spectrometers of low resolving power are used in atomic absorption spectrometry. With them the determination of a single isotope may be possible if the primary source emits the spectrum of the desired isotope, but not of another interfering isotope of the same element.

If only hollow cathodes are available, each with several isotopes and with different isotopic compositions, the determination of isotopic concentrations or ratios can be done in another way. Suitable methods for similar problems are well-known in molecular spectroscopy as "multicomponent analysis"; considerable theoretical treatment is necessary. Goleb (11, 12) has developed such a technique for the determination of the isotopic composition $^{235}U/^{238}U$ by atomic absorption spectrometry. He described two modifications:

(1) In the absorption tube method the sample of the unknown isotopic composition is placed in the absorption tube. To determine the ratio of the two isotopes, two emission tubes as radiative sources with different isotopic compositions are needed. The cathode material of one emission

source was 99.3% ^{238}U and 0.7% ^{235}U, while the material of the other was 7% ^{238}U and 93% ^{235}U. With the two emission sources a different absorption signal is measured because the absorption depends on the isotopic composition of the sample in the absorption tube.

(2) In the emission tube method the emission spectrum of the sample is excited in an emission hollow cathode used as the primary source. The cathode material of the second emission tube and of the absorption tube was 94.6% ^{235}U and 5.4% ^{238}U. A different absorption signal is measured with both emission tubes except when the isotopic composition in both

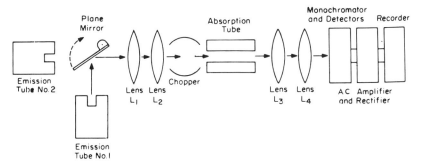

Fig. 3. Schematic diagram of the apparatus used for the determination of uranium isotopes. [Reprinted from Ref. (*12*) by courtesy of Elsevier Publishing Company.]

emission tubes is identical. A schematic diagram depicting the apparatus is shown in Fig. 3.

Analytical lines for the isotopic analysis of uranium are at 5915, 4153, and 5027 Å. If ^{238}U were to be determined in ^{234}U or ^{236}U, the 234, 235, and 236 isotopic components of the 4153-Å uranium line apparently do not interfere. The samples analyzed were in the form of metals, oxides, and salts. By this technique the concentration of uranium isotopes were reported with high precision and accuracy.

Another example of the determination of isotopic compositions by the hollow-cathode technique is the determination of ^{6}Li and ^{7}Li (*13*). Water-cooled hollow cathodes were used as emission and as absorption sources. The cathode cylinders were made of copper. To obtain the maximum absorption for the lithium resonance lines at 6708 Å, both tubes were operated at the lowest helium pressure possible, about 2 Torr. In the emission hollow cathode ^{6}Li and ^{7}Li fluoride compounds, prepared from ^{6}Li and ^{7}Li hydroxide solutions, were inserted. About 10 mg of lithium fluoride in the emission hollow cathode was sufficient to produce a stable emission lasting 2–3 days after the sample was sputtered on the walls of the cathode for 1–2 hr.

The isotopic lithium compounds used to prepare standards for the absorption hollow cathode were made by dissolving high purity ^{6}Li and ^{7}Li metals in distilled water. Preparation of standards and samples followed a detailed procedure. The drying technique used to achieve a uniform deposit on the walls of the hollow-cathode cylinder is critical.

C. THE GRAPHITE CELL

Work of Woodriff and Ramelow (14) has shown that the use of systems involving graphite cells makes it possible to achieve high sensitivity. The graphite cells are patterned after the carbon resistance furnace

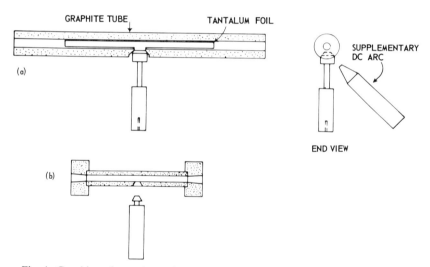

Fig. 4. Graphite tube as absorption cell used by L'vov: (a) 1959 version and (b) 1967 version.

of King (15). L'vov (16–18) volatilized the sample into a heated tube with a carbon arc. Figure 4a shows a diagram of the first version of his graphite cell, an absorption tube 10 cm in length and open at both ends. A tantalum foil, placed inside the tube, diminished the loss of the sample vapor by diffusion through the wall of the graphite tube. The sample, as a solution, is placed on the end of a carbon electrode and evaporated to dryness. To prevent the sample solution from soaking into the carbon, the tip of the electrode is first coated with a solution of polystyrene in benzene. The electrode carrying the sample is introduced into a conical opening in the middle of the graphite tube. The tube is heated by an ac current supplied by a 10-kW transformer. Simultaneously the electrode is heated externally by a dc carbon arc. The furnace is placed in an argon chamber which is hermetically sealed by an aluminum cover and rubber gaskets. Two quartz windows permit light to pass through the absorption tube. The electrodes are changed by a mechanism which allows different samples to be analyzed without opening the argon chamber. The time required for the evaporation of the sample can be 0.1 sec; therefore, the loss of sample vapor by diffusion is small. Under these conditions the absorption can be measured with high sensitivity (*19–22*).

In a revised version of the graphite cell by L'vov and Lebedev (*23*) the tube was shorter and no tantalum foil was placed inside the tube. To diminish diffusion of the vapor through the wall, it was coated with pyrolytic graphite or else fabricated entirely from pyrolytic graphite. Another difference involves the method of heating since heating by an external arc is inefficient as only a very small fraction of the power is used to heat the sample. A more efficient method involves heating the head of the electrode by an arc discharge between the electrode and the tube. By this method the heat is released mainly in the head of the electrode causing the sample to evaporate rapidly.

A somewhat simpler graphite tube furnace is shown in Fig. 5. Here Massmann (*24, 25*) places the sample in the tube and then heats the tube. The absorption cell is the graphite tube in the argon chamber. A slow stream of argon flows through the chamber to inhibit the burning of the graphite tube. The absorption tube is 55 mm in length, inner diameter is 6.5 mm, and wall thickness is 1.5 mm. A hole, 2 mm in diameter, is drilled in the wall at the middle of the tube to permit the introduction of the sample solutions. Solid samples are inserted from one side of the tube. In comparison with the L'vov design, the samples are not inserted into the tube by a special electrode mechanism but directly into the tube, and the evaporation of the sample is accomplished by electrical resistance heating of the tube wall and not by a supplementary arc. After the sample is introduced, the cell is heated in a few seconds up to 2600°C by a current up to 500 A. Since the samples are completely evaporated during the heating period, the cell is ready for the next sample after a cooling time of 20–30 sec. Depending on the maximum temperature required, the lifetime of a graphite cell ranges from 200–300 analyses.

Atomic absorption with the graphite cells described are for analytical

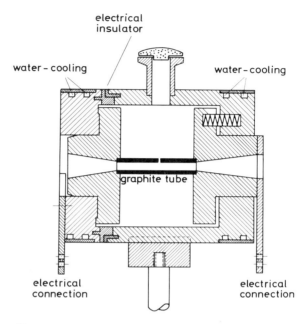

Fig. 5. Graphite tube in a water-cooled argon chamber.

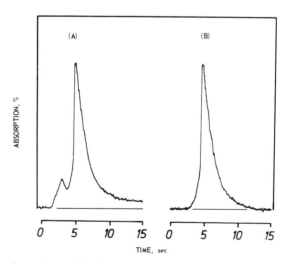

Fig. 6. Time dependence of the absorption of zinc: (A) without and (B) with reference signal.

procedures with a restricted amount of sample. Therefore, during the heating of the cell a time-dependent signal is always observed. Figure 6 shows two responses in the determination of zinc. Both responses are due to absorption but the technique of measurement was different. Figure 6A shows the time-dependent absorption at the wavelength of the zinc line at 2139 Å without background correction. The signal is usually the same whether or not one modulates the light from the primary source. In addition to the absorption of zinc, an interference due to background is observed. This background absorption may be of different magnitude and may have a different time dependence, depending upon the sample matrix in which the element is determined (26). Interference from background absorption can be minimized by simultaneously measuring the intensity of a nearby nonabsorbing line. For this purpose a spectrometer with dual detector channels is required. As shown in Fig. 6B, the time dependence of the intensity ratio of the zinc resonance line and a nonabsorbing argon line eliminates the background absorption. Another way to correct for background interference is to measure simultaneously a suitable reference signal using two light sources, as proposed by Koirtyohann and Pickett (27).

1. Sensitivity

The sensitivity of atomic absorption analysis with graphite cells depends on the geometry of the graphite cell. L'vov (28) has shown that the sensitivity is inversely proportional to the square of the inner diameter of the graphite tube if small amounts of sample are analyzed. In order to compare the sensitivities measured with graphite cells of different geometry, it is useful to calculate the sensitivity in respect to a special inner diameter of the tube. A comparison of sensitivities measured by different authors and with graphite cells of different geometry and by different methods of heating is shown in Table 2. Sensitivity is defined as the concentration, in grams, necessary to obtain an absorption of 1%. The data given in the table are valid for measurements with a graphite tube of 6.5-mm inner diameter. If the measurements had been made with tubes of another diameter, the data would have to be adjusted to this diameter. The difference in the sensitivity obtained by each method may be explained by the different velocity of the evaporation process. Since L'vov used a supplementary arc, the evaporation may be faster with an increase in absorption.

2. Detection Limits

More important than the sensitivity is the limit of detection used to characterize an analytical method. The detection limit depends on the

TABLE 2

SENSITIVITIES OBTAINED BY ATOMIC ABSORPTION WITH
GRAPHITE CELLS

		Sensitivity, wt/1 % absorption	
Element	Wavelength, Å	L'vov furnace	Massmann furnace
Ag	3281		0.8 pg
Al	3093	7.7 pg	
As	1890		0.2 ng
Ba	5535	41 pg	
Bi	3068	26 pg^a	0.2 ng
Cd	2288	0.5 pg	2.0 pg
Co	3527	200 pg^a	
Cr	4254	41 pg	
Cs	4555	3.3 ng^a	
Cu	3248	1.8 pg	9 pg
Fe	2967/2483	17 pg	20 pg
Hg	2537	170 pg^a	200 pg
In	3039	3.0 pg	200 pg
K	4044	960 pg	
Li	6708	20 pg	
Mg	2852		0.5 pg
Mn	2795	0.5 pg^a	8 pg
Mo	3133	11 pg^a	
Na	5890		6 pg
Ni	3525	180 pg^a	70 pg
Pb	2833	15 pg^a	10 pg
Rb	4202	1.5 ng^a	
Sb	2311	16 pg^a	0.1 ng
Se	1961		1.5 ng
Sr	4607	8.2 pg	10 pg
Te	2259	290 pg^a	
Tl	2768		40 pg
Ti	4982	2.0 ng	
Zn	2139	0.2 pg^a	0.8 pg

^{a} Analyses were made at a pressure between 3 and 6 atm (28).

sensitivity and on the noise. If a graphite tube with an inner diameter of 6.5 mm and a length of 50–60 mm is used, one can measure 1 % absorption of an analyte with relatively small amounts of sample. This means that the total sample (in grams) given for 1 % absorption is also virtually the detection limit. These detection limits are valid for single-channel measurements, with the only exceptions being the determinations of arsenic and selenium

(29). In the wavelength region below 2000 Å, serious interferences can be avoided if the intensity of a nonabsorbing line is simultaneously measured. Nevertheless the lowest absorption that can be reliably detected for these two elements is relatively high, about 3%.

If primary sources with high radiation intensities are available, such as high brightness lamps, it is possible to reduce the diameter of the graphite tube without incurring a higher noise. In this case the detection limit can be lowered considerably because of the higher sensitivity caused by the improved geometry of the absorption volume (30, 31).

3. Precision

The precision for atomic absorption with graphite tubes is less than that obtained with flames from a laminar flow burner. The relative standard deviation for the graphite tube is 4–12% if (1) the solid matrix in the sample solution is 2 g per 100 ml or less and (2) the sample does not exceed 100 μl volume. An increase in precision is attained if an internal standard is used and the absorbance of the reference element is measured simultaneously with the analyte. This technique eliminates variations due to the evaporation process in the cell and gives a standard deviation of 3–7%.

4. Accuracy

The accuracy is closely related to the influence of the matrix. Measuring the maximum intensity of the analytical signal is only one of several possible approaches. Another method involves integration of the signal (31). Often when the matrix concentration is changed, not only is there a change in the analytical signal but also remarkable changes in the curve form. In such instances less influence of the matrix and often better accuracy can be attained by measuring the integral of the absorbance.

III. Other Vaporization Methods

A. Heated Carbon Filament

An electrically heated carbon filament was used as a source of atoms by Anderson et al. (32). They placed a 1–5 μl sample on a carbon filament, 2 mm in diameter and 4 cm in length, mounted horizontally between two stainless steel supports and within a detachable glass cell through which a continuous stream of argon flowed. The liquid was allowed to evaporate. Then for about 3 sec a current of 100 A at 7 V was passed through the filament which acquired a temperature of about 2500°C. A cloud of atomic vapor was produced within the cell which interacts with the light beam from the source passing longitudinally above the filament in atomic absorption, or at right angles to it in atomic fluorescence. The spectrometer slit lay along the axis of the filament. A peak appeared on the recorder chart, and the height of this peak was plotted against the absolute mass of the metal in the original standards applied to the filament. Working curves appear linear over a wide range of concentrations. Individual readings were reproducible to within 5%; sensitivity extends to picogram amounts of metals. The limiting factor in reproducibility was the application of the sample to the filament. This vaporization method offers freedom from interference by refractory oxide forming elements due to the reducing conditions of the filament and the complete absence of oxygen.

B. RADIATION HEATING

To evaporate samples for atomic absorption by radiation heating, two different types of radiation sources have been reported; a pulse discharge unit by Nelson and Kuebler (*33, 34*), and a solid state laser by Mossotti et al. (*35*). Solid samples can be evaporated with both sources by a radiation pulse of short duration. The main difference between the two methods is that in the pulse discharge no optical arrangement was used to focus the radiation onto the sample, whereas with laser heating the evaporation was done by a focused laser beam.

In the pulse discharge unit, solid samples must be in the form of several fine filaments or thin strips. Salts can be deposited from solution on an array of thin graphite strips and dried in vacuo. To achieve a high irradiance on the sample strips, a transmitting cell with the strip array was placed in the interior of a helical-wound flash lamp, with nitrogen, helium, or argon flowing through the cell at atmospheric pressure. The duration of the flash was about 3 msec. Because of the pulse evaporation of the sample, the measurement must be done in a short time. Therefore, a second flash lamp is used as a primary source. The source flash persisted for 20 μsec and could be ignited with variable delays up to 6 msec after the start of the heating flash. The absorption spectra were recorded photographically, but no attempt was made to measure the absorption quantitatively. Range was 10–50 μg/ml.

If a pulse of focused laser light is directed onto the analytical specimen, samples can be analyzed directly from the solid state. Mossotti et al. (*35*) used a ruby laser with a Q-switch. The laser beam was brought into a sharp focus on the sample surface by a lens system of 15-mm focal length. The craters produced averaged 200 μm in depth and in diameter. As a primary source a low pressure flash lamp was used; the light pulse being triggered by the rotating Q-switch. A 2-m grating spectrometer was used, and the analytical and reference signal were simultaneously recorded with a time constant of 15 μsec by an oscilloscope. Figure 7 shows the time distribution

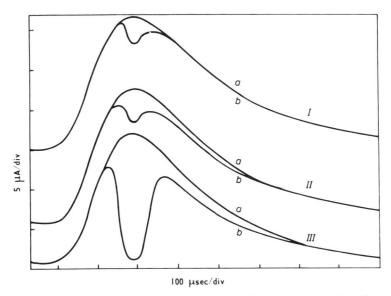

Fig. 7. Time dependence of the photocurrent during laser evaporation of a solid sample: (I) zeroth order; (II) background next to Cu 3247 Å; (III) Cu 3247 Å; (a) no target and (b) copper target. [Reprinted from Ref. (*35*) by courtesy of Pergamon Press.]

of the intensity measured in three different photoelectric channels. The discharge of the primary source was initiated 100 μsec after the start of the oscilloscope sweep, and the laser was triggered 155 μsec after the initiation of the primary source. In addition to the resonance absorption an unspecific background absorption occurs. To diminish such interference it is necessary to measure simultaneously the background absorption in another photoelectrical channel. Under these conditions the lowest measurable absorption is nearly 10%. For the determination of traces in graphite, the lowest detectable concentration is 25 μg/ml for calcium, 40 μg/ml for silver, and 35 μg/ml for copper. Since only a small amount of sample is vaporized by the laser beam, the detection limits are extremely low and in the same order as those reported with graphite cells.

REFERENCES

1. B. J. Russell and A. Walsh, *Spectrochim. Acta*, **15**, 883 (1959).
2. B. M. Gatehouse and A. Walsh, *Spectrochim. Acta*, **16**, 602 (1960).
3. A. Walsh, *Spectroscopy*, The Institute of Petroleum, London, 1962, pp. 13–26.
4. A. Walsh, *Proceedings Xth Colloquium Spectroscopicum Internationale*, Spartan Books, Washington, D.C., 1963, pp. 127–142.
5. N. P. Ivanov, M. N. Gusinski, and A. D. Esikov, *Zhur. Anal. Khim.*, **20**, 1133 (1965).
6. H. Massmann, *Spectrochim. Acta*, **23B**, 393 (1970).
7. J. A. Goleb and J. K. Brody, *Anal. Chim. Acta*, **28**, 457 (1963).
8. K. R. Osborn and H. E. Gunning, *J. Opt. Soc. Am.*, **45**, 552 (1955).
9. B. V. L'vov and B. I. Mositchev, *Zhur. Prikl. Spektrosk.*, **4**, 491 (1966).
10. O. P. Botchkova and E. Ja. Shreider, *Spectral Analysis of Gas Mixtures*, Fizmatgiz, Moscow, 1963 (in Russian).
11. J. A. Goleb, *Anal. Chem.*, **35**, 1978 (1963).
12. J. A. Goleb, *Anal. Chim. Acta*, **34**, 135 (1966).
13. J. A. Goleb and Yu Yokoyama, *Anal. Chim. Acta*, **30**, 213 (1964).
14. R. Woodriff and G. Ramelow, *Spectrochim. Acta*, **23B**, 665 (1968).
15. A. S. King, *Astrophys.*, **28**, 300 (1908).
16. B. V. L'vov, *Ing Fiz. Zhur.*, **11**, 44 (1959).
17. B. V. L'vov, *Ing. Fiz. Zhur.*, **11**, 56 (1959).
18. B. V. L'vov, *Spectrochim. Acta*, **17**, 761 (1961).
19. G. I. Nikolaev and V. B. Aleskovsky, *Zhur. Anal. Khim.*, **18**, 816 (1963).
20. G. I. Nikolaev, *Zhur. Anal. Khim.*, **19**, 63 (1964).
21. G. I. Nikolaev and V. B. Aleskovsky, *Zhur. techn. Fiz.*, **24**, 753 (1964).
22. G. I. Nikolaev, *Zhur. Anal. Khim.*, **20**, 445 (1965).
23. B. V. L'vov and G. G. Lebedev, *Zhur. Prikl. Spektrosk.*, **7**, 264 (1967).
24. H. Massmann, *Proceedings XIIth Colloquium Spectroscopicum Internationale*, Hilger and Watts, London, 1965, pp. 275–278.
25. H. Massmann, *Spectrochim. Acta*, **23B**, 215 (1968).
26. H. Massmann, *Reinststoffanalytik*, Part 2, Akademie-Verlag, Berlin, 1965, pp. 297–308.
27. S. R. Koirtyohann and E. E. Pickett, *Anal. Chem.*, **37**, 601 (1965).
28. B. V. L'vov, *Atomic Absorption Spectral Analysis*, Nauka, Moscow, 1966.
29. H. Massmann, *Z. anal. Chem.*, **225**, 203 (1967).
30. B. V. L'vov, *Zhur. Prikl. Spektrosk.*, **8**, 517 (1968).
31. H. Massmann, *Methodes Physique d'Analyse*, **4**, 193 (1968).
32. R. G. Anderson, I. S. Maines, and T. S. West, *International Atomic Absorption Spectroscopy Conference*, Sheffield, 1969, paper A-7, Hilger and Watts, London, 1969.
33. L. S. Nelson and N. A. Kuebler, *Proceedings Xth Colloquium Spectroscopicum Internationale*, Spartan Books, Washington, D.C., 1963, pp. 83–90.
34. L. S. Nelson and N. A. Kuebler, *Spectrochim. Acta*, **19**, 781 (1963).
35. V. G. Mossotti, K. Laqua, and W. D. Hagenah, *Spectrochim. Acta*, **23B**, 197 (1967).

5 The Optical Train

John A. Dean

DEPARTMENT OF CHEMISTRY
UNIVERSITY OF TENNESSEE
KNOXVILLE, TENNESSEE

The function of the optical train differs somewhat for the different modes: flame emission and atomic absorption. However, the general arrangement of an atomic absorption spectrometer is no different from a flame emission spectrometer except for the addition of a light source and possibly a chopper.

In flame emission the task of the optical train is to collect the emitted light, render it monochromatic, and then focus it onto the surface of the photoreceptor. Since only a small fraction of the total radiant flux emitted by the flame enters the flame spectrometer, a concave mirror is frequently placed behind the flame, with its center of curvature in the flame. In this way the intensity of the emitted light is nearly doubled. The optical beam is generally chosen as to avoid the flickering tip and unsteady edges of the flame as well as the bright inner cone.

In atomic absorption spectrometry the optical train has two specific functions: to reject all lines of other elements, including those of the filler gas in the light source, and to prevent the photoreceptor from being overloaded with light.

How the specific functions are achieved by various wavelength isolators, how well monochromators perform in practice, and the various arrangements of components in optical trains are the subject materials of this chapter. The useful wavelength range of atomic absorption and flame emission methods currently runs between the arsenic line at 1937 Å and the cesium line at 8521 Å.

I. Wavelength Isolator

Although requirements for the wavelength selector vary considerably from one element to another, basically it must isolate effectively the radiation due to the test element while eliminating adjacent lines. Additionally, in the emission mode one must be able to select the most favorable ratio between background and analytical line radiation.

Less expensive instruments may use an interference filter to isolate the radiation characteristic of a given element. Better isolation of spectral energy can be obtained with a prism or grating as dispersing medium. These dispersing devices, in conjunction with entrance and exit slits, suitable baffles, and mirrors, form a monochromator.

A. FILTERS

The tinted glass absorption filter consists of a solid sheet of glass that has been colored by a pigment which is either dissolved or dispersed in the glass. Composite filters are constructed from sets of unit filters. One series consists of sharp cutoff filters that pass long wavelengths, the red and yellow series, and the other comprises long wavelength cutoff filters, the blue and green series. The transmittance of the individual cutoff filters

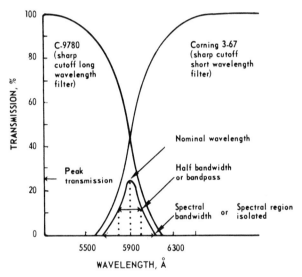

Fig. 1. Spectral transmittance characteristics of an absorption glass filter and its component cutoff filters.

and the combined transmittance for a filter with a nominal wavelength at 5900 Å are shown in Fig. 1. Glass filters have a wide bandpass of 350 to 500 Å at one-half maximum transmittance, and their peak transmittance is only 5–20%, decreasing with improved spectral isolation. Combination filters are seldom employed now in flame photometers although they enjoyed prominence in early photometers. However, cutoff filters enjoy wide use as blocking filters to suppress unwanted spectral orders from gratings and interference filters.

Interference filters employ thin metallic or dielectric layers to produce interference phenomena at desired wavelengths, thus permitting rejection of unwanted radiation by selective reflection. Their construction follows. A semitransparent metal film is deposited on a plate of glass. Next, a thin layer of some dielectric material, such as SiO or MgF_2 is evaporated on top of this, and then the dielectric layer is in turn coated with a film of metal. Finally another plate of glass is placed over the films for mechanical protection. The completed filter is shown in Fig. 2. A portion of the light incident upon the face of the filter is reflected back and forth between the metal films. Constructive interference between the different pairs of superposed light rays occurs only when the path difference is exactly one

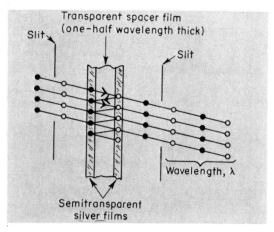

Fig. 2. Path of radiation through an interference filter.

wavelength or a multiple thereof. Since the path difference is now in the dielectric of refractive index n, the wavelengths of maximum transmission for normal incidence are given by

$$\lambda = 2nb/m \tag{1}$$

where b is the thickness of the dielectric spacer and m is the order number. For example, a dielectric layer of $n = 1.35$ that is 185 nm thick will provide a first-order filter of 5000 Å peak wavelength. This filter also passes bands centered at 2500 Å and 1670 Å. Unwanted transmission bands can be eliminated by using an appropriate sharp cutoff absorption filter for the protecting glass cover, as indicated in Fig. 3. Interference filters have a bandpass of 100–150 Å and peak transmittance of 40–60%.

Multilayer interference filters consist of layers of nonabsorbing material of alternately high and low indices of refraction deposited on an optical base, followed by a transparent spacer film, and then by more layers of similar material. These filters have transmissions of 50–70% and bandwidths of 10–50 Å.

By depositing a wedge-shaped layer of dielectric between the semi-reflecting metallic layers, a continuously variable transmission filter is obtained. At each point along the base of this filter a different wavelength band will be transmitted. In use the wedge is moved past a slit to select wavelengths.

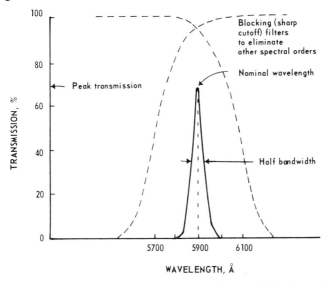

Fig. 3. Transmission of an interference filter and its cutoff (blocking) filters.

The efficiency of a filter can be characterized by the ratio of the desired transmittance at the line or band to the residual undesired transmittance in regions of the spectrum where the filter should be entirely opaque. Consider, for example, a solution containing 1000 μg/ml of sodium which gave a signal of 0.5 division through a potassium filter. Through the same filter 10 μg/ml of potassium gave a deflection of 45 divisions, so that 0.11 μg/ml of potassium would have given a deflection of 0.5 division, an efficiency of 98.9%.

B. PRISMS

The action of a prism depends on the refraction of light by the prism material. The dispersive power $(dn/d\lambda)$ depends on the variation of the index of refraction with wavelength. A light ray entering a prism at an angle of incidence i will be bent toward the normal (vertical to the prism face) and, at the prism-air interface, it is bent away from the vertical (see Fig. 7). The image of the entrance slit is projected onto the exit slit as a series of images ranged next to each other, caused by light of shorter wavelengths being more strongly bent than light of longer wavelengths. A nonlinear wavelength scale results. Flint glass provides about threefold better dispersion than quartz, and is the material of choice for the near infrared and the visible region of the spectrum. Natural quartz or fused silica is required for work in the ultraviolet region.

C. Gratings

A grating consists of a large number of parallel, equally spaced grooves ruled upon a metal surface, usually aluminum. Replica grating fabrication

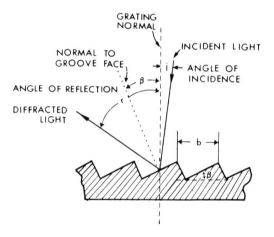

Fig. 4. Cross-section diagram of a diffraction grating showing the "angles" of a single groove which are microscopic in size on an actual grating. Symbols are: i = angle of incidence, r = angle of reflectance, β = blaze angle, and b = grating spacing.

is described by Willard et al. (*1*). When light strikes a diffraction grating, each wavelength is reflected in a predictable direction (Fig. 4). The condition of constructive interference is stated in the grating formula:

$$b(\sin i \pm \sin r) = 1\lambda = 2\lambda = 3\lambda = m\lambda \qquad (2)$$

where b is the distance between adjacent grooves, i is the angle of incidence, r is the angle of reflectance, and m is the order number. A positive sign applies where incoming and emergent beams are on the same side of the grating normal. To illustrate the use of Eq. (2), the primary angle at which light of 3000 Å will be diffracted at normal incidence ($i = 0°$), by a grating ruled 1180 grooves/mm, in the first order, is given by

$$\sin r = m\lambda/b - \sin i$$
$$= [1 \times 3000 \times 10^{-8} \text{ cm}/(1/11{,}800) \text{ cm}] - 0$$
$$= 0.254$$

The angle having this sine is 14.7°.

In each angular direction of reflection, wavelengths of other orders also occur (Fig. 5). For example, the magnesium line at 2852 Å would also appear at 5704 Å in the second order, at 8556 Å in the third order, and so on. Use of the second-order line at 5704 Å offers an operator twice the dispersion obtainable in the first order (q.v.); it would also circumvent the need for windows of quartz or fused silica. Of course, one has to be alert to the possibility that a higher-order line of one element might provide spectral interference for the first-order line of another element.

Viewing the grating on the normal results in linear dispersion and a linear scale for wavelength equal to m/b. The smaller the b value, the more widely spread will be the spectrum. The second-order spectrum has twice the linear dispersion of the first order, the third three times, and so on. Most emission and absorption flame spectrometry is done using first and second orders.

Two factors limit the number of grooves per millimeter that can be used for a given wavelength. Neither i nor r can be greater than about 65°, and the groove aspect as seen by the light cannot be substantially less than the wavelength of the light, or the grating acts as a mirror reflecting light rather than dispersing it. Standard gratings will have approximately 600, 1200, or 1800 grooves/mm ruled on an area varying from 25 × 25 mm to 102 × 102 mm. The ruled area of a grating should be large enough to intercept all the incident light even when the grating is turned to its extreme angular position. Any smaller area will decrease the useful light in the spectrum and increase that going into the zero order, or wasted by missing the grating altogether.

When the grating is ruled, the groove or blaze angle β can be adjusted so that most of the light intensity will be concentrated in the wavelength region of greatest interest. This is called *blazing* the grating. The blaze wavelength λ_β is defined as that wavelength for which the angle of reflectance from the groove face and the angle of reflection from the grating

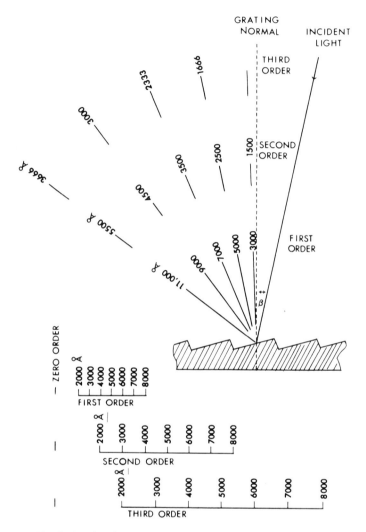

Fig. 5. Overlapping orders of spectra from a reflection grating.

normal are identical; then also the angle of incidence and the angle of reflection are equal; that is, when $i = r$, then $\beta = i$. In this case the grating equation reduces to

$$m\lambda_\beta = 2b \sin \beta \tag{3}$$

For other angular relationships, the blaze angle is

$$\beta = (i \pm r)/2 \tag{4}$$

The spectral region covered corresponding to the half-intensity range of a grating in its first order extends approximately from two-thirds its blaze wavelength to twice the blaze wavelength. In the second and higher orders the grating intensity curve tends to be more balanced and extends from $\frac{2}{3}m\lambda_\beta$ to $\frac{3}{2}m\lambda_\beta$.

D. Monochromator Designs

The two most popular monochromator designs are the Ebert (or a modification) and the Littrow. The merits of the two systems have been

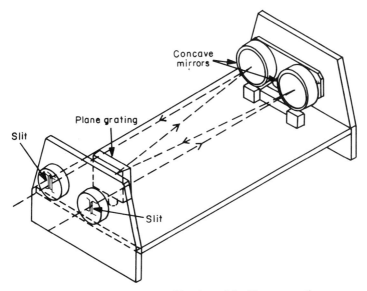

Fig. 6. Czerny-Turner modification of the Ebert mounting.

compared for many years. There seems to be little to choose between them. Both are compact designs.

The Ebert mounting (2) is a compact arrangement using a plane grating (Fig. 6). Focusing is achieved by off-center reflections from a large concave mirror which collimates the entrant light striking the grating and intercepts the dispersed beam and focuses it on the exit slit. In the side-by-side Czerny-Turner mounting (3), two smaller concave mirrors replace the single large mirror. These mountings are stigmatic and also achromatic, so that the rays of all wavelengths are brought to focus at the exit slit without

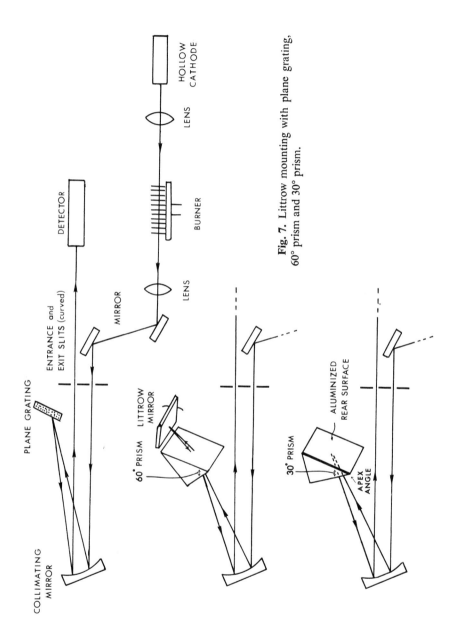

Fig. 7. Littrow mounting with plane grating, 60° prism and 30° prism.

changing the slit-to-mirror distance. Entrance and exit beams are stationary. Wavelength is readily adjusted without affecting focus by rotating the grating on its axis. A sine-bar drive produces a direct wavelength readout (*q.v.*). The Czerny-Turner system provides cancellation of aberrations and the capability for balancing types of aberrations to optimize performance for two-dimensional precision entrance and exit slits. Fastie (*4*) suggested an "under-over" design in which the entrant rays pass below the grating while the emergent rays pass above.

The Littrow mounting, shown in Fig. 7, will accommodate a 30° quartz prism backed by an aluminized surface, a 60° prism and separate Littrow mirror, or a grating. The image of the entrance slit is collimated by a spherical mirror and directed onto the dispersing device. The refracted or diffracted beam is then sent back to the same collimating mirror at a different height and the collimated beam is then projected and focused onto the exit slit, which selects a portion of the dispersed spectrum for transmission to the photoreceptor. The upper and lower parts of the same slit assembly are used as entrance and exit slits, providing perfect correspondence of slit widths. The slit system must be continuously adjustable with prisms but may be one of several fixed openings when a grating is the dispersing device. In the 30° prism arrangement, the prism (or the grating when it is used as the dispersing device) is rotated by means of a mount connected to the wavelength scroll, whereas for the 60° prism the separate Littrow mirror is turned through a small angle to obtain different wavelengths. With prisms the wavelength scale must be calibrated relative to the refractive index of the prism material.

E. ECHELLE-PRISM COMBINATION

A combination of an echelle grating with a prism provides extremely high resolution (*5*). Typically an echelle is ruled 73–78 grooves/mm over an area of 75 × 150 mm and is blazed at about 63°. The spacing between rulings is equivalent to approximately 50 to 100 times the wavelength of the light to be dispersed. The echelle is used in its 50th order at 5000 Å and 100th order at 2500 Å, for example. In the prism-echelle arrangement (Fig. 8), the quartz prism is mounted in the optical path before the echelle. The prism diffracted spectrum is dispersed horizontally by the echelle. The many orders, each covering only a small spectral range, are separated and spaced evenly in the vertical direction by the prism. Dispersion by the prism varies monotonically with wavelength at a rate which very nearly

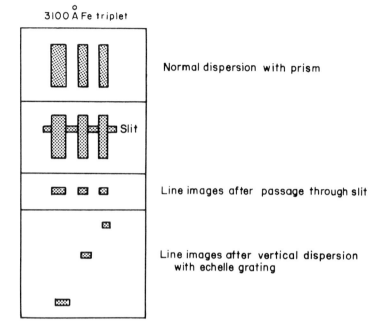

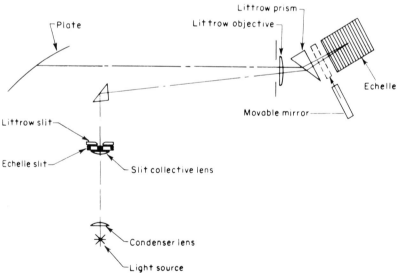

Fig. 8. Prism-echelle arrangement and how the dispersion is achieved. (Courtesy of Bausch & Lamb Optical Co.)

maintains constant spacing between adjacent grating orders. The result is a spectral display that is equivalent to a spectrum spread over 1.6 m. An echelle can be treated as a very coarse grating to compute dispersion, resolution, and optical speed.

F. ORDER SORTER

In a prism-grating double monochromator designed to operate in the grating second order, the prism monochromator is intended to operate

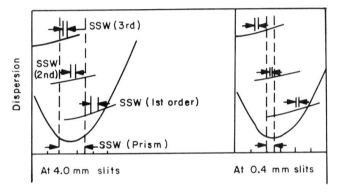

Fig. 9. Solid curves illustrate dispersion curves for the prism and three orders of a grating. Spacings on vertical lines indicate spectral slit width (SSW) passed by each monochromator at 4.0- and 0.4-mm slits.

essentially as an order sorter. The second-order grating radiation is allowed to pass while the first, third, and higher orders are rejected. To accomplish this rejection the bandpass of the prism monochromator must be sufficiently narrow to include only radiation within the spectral slit width (twice the bandpass) from the second grating order (see Fig. 9). A simpler arrangement places a fore-prism between the source and the entrance slit of the monochromator; this serves to place the various orders of spectra one above the other.

II. Auxiliary Optical Components

In this section we will discuss optical components other than the wavelength selector that may be found in a spectrometer, and outline their functions. In flame emission spectrometry, the optical system must concentrate light from the flame into a narrow, well-defined image that can serve as an apparent source. A collimating lens or mirror must then collimate the radiation spreading from the apparent source—the entrance slit of the monochromator. Following diffraction or dispersion, the dispersed or diffracted light must finally be brought to a focus on the exit slit. Lens optics have the unavoidable disadvantage that the instrument can be used only over the limited spectral range for which the lens material has the required transmission. Mirrors have practically the same efficiency in all spectral regions. Parabolic mirrors are free of chromatic aberrations, which arise in lenses because different wavelengths of light focus at different distances from a single lens due to the variation of the refractive index with wavelength.

A. POSITIONING THE SOURCE

For rapid interchangeability between hollow-cathode lamps, particularly those used routinely, several arrangements are used. The lamps can be arranged normal to the main optical axis (Chapter 9, Fig. 10). A mirror-lens combination rotates from one detent position to another to pick up light from the desired lamp and direct it down the long axis of the flame. The first lens images the bright cathode spot in the center of the flame. The second lens relays the cathode image to the entrance slit of the monochromator.

The turret-type mount accommodates a number of lamps at once. Indexing of the turret knob brings a new lamp into electrical contact with the power supply and automatically aligns the lamp with the optical axis of the system. Lamps may also be positioned on a horizontal lamp bank. As required the individual lamp is indexed into the optical axis. Additional lamp banks may be substituted quickly on the carrier.

B. REFLECTION DEVICES

Mirrors are made by adding a reflecting metal coating to an optical flat or a concave configuration. Metals used include aluminum, gold, silver, and rhodium. Front-surface mirrors avoid image distortion that occurs from the multiple reflections from a back-surface mirror, but must be protected from deterioration and abrasion.

Total internal reflection is sometimes preferred over metallic reflection of mirrors to avoid losses in intensity. Total internal reflection occurs when light is incident on the surface of a medium whose refractive index is smaller than that of the medium (air) in which the radiation is traveling. Also the radiation must strike at an angle greater than the critical angle for the particular interface. Now the critical angle of an air-flint glass interface is 42°. This makes possible the use of 45° prisms as totally reflecting surfaces (Fig. 10).

For the dispersing device to function efficiently, the light must be collimated in a parallel beam. In a monochromator the collimating

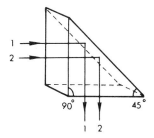

Fig. 10. Totally reflecting prism.

mirror is positioned so that the slit is located at the focal length of the collimator. The diameter of the collimator is chosen as required to illuminate completely the entire area of the dispersing device. Upon returning from the dispersing device, the beam is still parallel, and so it is refocused by the collimator on the exit slit as a series of images of the entrance slit, again at the focal length of the collimator. The reader is referred to the optical diagrams of the Ebert, Czerny-Turner, and Littrow mountings (Figs. 6 and 7) to observe the function of the collimator.

Whenever radiation is incident upon a boundary between dielectrics across which there is a change in refractive index, reflection occurs. The greater the difference in refractive indexes, the larger is the reflectance. Conversely, if $n_1 = n_2$, there can be no reflection. For optical components that must transmit light, the phenomenon of interference is utilized to produce a low reflection coating, generally SiO in the visible and MgF_2 in the ultraviolet. The refractive index of the vacuum-deposited coating should be equal to $(n_1 n_2)^{1/2}$, where n_1 is the refractive index of the optical component and n_2 that of air. Equal quantities of light will be reflected from the outer surface of the coating and from the boundary surface between it and the glass (or other transmitting material). When the coating thickness is one-fourth the wavelength in the center of the spectral band whose transmission is desired, the light reflected from the first surface will be 180° out of phase with that reflected from the second. Complete destructive interference will result. In this manner the overall reflection from a surface can be reduced from 4–5 % to less than 1 %. This treatment is extremely effective in eliminating stray reflected light. Even so, each mirror surface at 3000 Å has a reflectivity of 0.92. A good quality quartz lens has a transmission at that wavelength of about 0.95. If six or seven reflecting surfaces are involved, as well as several lenses, then energy losses are not at all trivial.

These film coatings are also used to improve reflection from mirrors and to protect the metal film on front surface mirrors. For this use the coating thickness is one-half the wavelength of the light. Reflectances can exceed 99.9 %.

C. BEAM SPLITTERS

Achromatic beam-splitting plates (Fig. 11) are used when an incident beam of light must be split spectrally. They are semitransparent mirrors

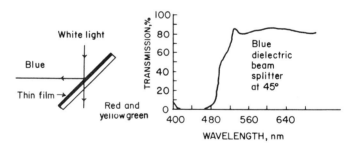

Fig. 11. Beam splitter.

on one surface of which is deposited a coating so that the reflection equals any desired value, the remainder of the incident light being transmitted. Also called dichroic mirrors, they are typically used at an angle of 45°. With white illumination the two separated beams are complementary. With discrete light sources the two separated beams may be utilized to isolate certain portions of the spectrum of the illumination source. Filters of this type are generally specified as to the color reflected.

D. Light Interrupter

When the light beam from the source is interrupted at a controlled frequency, only this modulated light can enter the electronics when the detector-amplifier circuit is tuned to this same frequency. The chopper is mounted between the hollow-cathode lamp mount and the flame in atomic absorption. For emission the chopper is positioned between the burner and the entrance slit of the monochromator. Electronic modulation is discussed in Chapter 6.

One type of chopper is a metal disk with two open opposite quadrants which is placed off-center in the light path. The remaining quadrants chop off the light source when the disk is rotated. In a double-beam system, the motor-driven chopper becomes a rotating sector mirror, which alternately passes the source light through the flame and past the flame. Mirrors are fastened on the source side of the closed quadrants of the metal disk.

The balanced vibrating members of a tuning fork permit elimination of the motor drive. The chopper is operated by an external transistor circuit. Signal voltage is available from the fork circuit as a reference. The tine amplitude is adjustable. The most useful type has the aperture at half-opening when at rest. At the outer swing of the tines, full aperture is achieved, and on the inward swing the vanes are closed and overlap slightly. In the half-and-half mode, the vanes are just in the meeting position at rest and at zero crossing. With this type the aperture width is about half of the first type.

E. Slits and Irises

Basically the purpose of the entrance slit is to provide a narrow source of light so that, after dispersion and refocusing in the plane of the exit slit, the amount of overlapping of the monochromatic images is limited. The exit slit then selects a narrow band of the dispersed spectrum for observation by the detector. In practice it is desirable to have the entrance and exit slits of equal width because it can be shown that, for conditions of a given resolving power, maximum radiant power is passed by the spectrometer when this is true. Under these conditions the width of a monochromatic image of the entrance slit is such that it will just be passed by the exit slit.

For perfect image formation, the monochromator must be equipped with slits with curved jaws where the radius of curvature equals one-half the distance between the entrance and exit slits. At full height they retain resolution while maximizing photoelectric light gathering power (*q.v.*). If the detector is somewhat shorter than the slit length, a condensing lens can be inserted after the exit slit to squeeze the beam appropriately so it conforms to the detector surface. Fixed slit openings, often 10, 25, 50, or 100 μm in width, are employed often with grating mounts. With prism monochromators the slit jaws are adjustable and operated by direct pressure on a steel spring. Although more expensive, this adjustable type is to be preferred even with grating mounts.

An adjustable iris diaphragm may be placed in the light path at a point in the beam where there is no imaging of the flame, as immediately before or behind the filter or one of the lenses. If set at about two-thirds open, the diaphragm can be adjusted to compensate for changes in sensitivity during prolonged operation or at different working periods. A calibrated iris diaphragm can serve as the readout scale; when operated by a logarithmic cam, the readings will be in absorbance units.

F. Multipass System

In principle it is possible to improve the sensitivity of an atomic absorption determination by passing the source radiation several times through the flame. Three- and five-pass systems are available; the latter is shown in Fig. 12. Since the light passes through the flame at a variety of heights

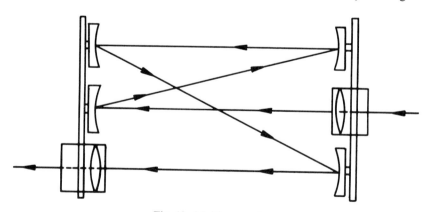

Fig. 12. Multipass optics.

and flame temperatures, including areas where the flame is not under as good control as in the optimum region, advantages are gained only when the region where useful absorption takes place extends over a considerable height and is not critically dependent on the air-fuel ratio. In this regard, Chapter 8 of Volume 1 should be consulted. From a five-pass system, approximately a threefold gain in sensistivity has been realized for an optimum situation.

G. WAVELENGTH READOUT

A linear wavelength readout is attained for the Czerny-Turner mounting as follows (Fig. 13): The grating G is rotated on its axis by arm GR,

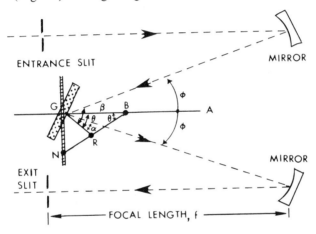

Fig. 13. Sine-bar wavelength drive for Czerny-Turner mount. [Reprinted from Ref. (6) by courtesy of Spex Industries, Inc.]

rigidly attached below the center of the grating face at a position perpendicular to the face. GN represents the precision lead screw along which N, its mating nut, rides while NB pivots around B, R, and N. B is a bushing along the centerline of the optical system GA, while GR turns the grating. Arms GR, NR, and RB are identical in length. A mechanical counter reading wavelength actually counts revolutions of the lead screw GN and therefore measures its effective length. Because NB is a fixed length, GN is proportional to $\sin \theta$. Since $\alpha = \theta - \phi$ and $\beta = \theta + \phi$,

$$m\lambda = b[\sin(\beta - \phi) + \sin(\theta + \phi)] \tag{5}$$
$$= b(\sin \theta \cos \phi - \cos \theta \sin \phi + \sin \theta \cos \phi + \cos \theta \sin \phi) \tag{6}$$
$$= 2b \sin \theta \cos \phi \tag{7}$$

Since cos ϕ is a constant for a constant deviation spectrometer and ϕ is 6°,

$$\lambda = k \sin \theta \tag{8}$$

and is therefore proportional to the length GN.

For recording purposes, the wavelength dial on the monochromator is replaced by a drive mechanism synchronized with the chart drive of a pen-recording potentiometer.

H. Optical Bench

Optical benches satisfy each need from the simplest prototype layout to the most exacting geometrical and physical research. In design they may be in the configuration of a rail in a V- or U-shape, or a flat bed about 12 cm in width (Chapter 9, Figs. 3 and 10). The latter is favored for precision units. Carriers are built on a modular form. The basic carrier base with various motions and accessory holders may be purchased or interchanged from carrier to carrier. The base of the carrier is held against the bench dovetail by a pressure screw. Carriers can be inserted or removed at any point along the bench. Often a vernier scale is attached to the base for determining the position of the carrier in relation to the bench. In this manner both horizontal and vertical travel is possible, with motion along a calibrated scale for reference purposes. With a bench and carriers, lamps, lenses, burners, and slits may be positioned experimentally for special purposes or their positions may be optimized for routine work.

III. Optical Qualities of Monochromators

The performance of a monochromator is expressed essentially by three factors: resolving power, speed or light-gathering power, and purity of light output. The resolving power depends on the dispersion and on the perfection of the image formation, while the purity is determined mainly by the amount of stray or scattered light. Large dispersion and high resolving power in monochromators will stress the importance of spectra with discrete lines, whereas continua and bands show up more clearly in the emission mode with instruments of small dispersion. It may be desirable to review the discussion on spectral interferences in Chapter 9 of Volume 1.

A. DISPERSION

Dispersion is a measure of the linear spread between two spectral lines, $dx/d\lambda$, in the plane of the exit slit of the monochromator. It is common practice to specify it as the reciprocal linear dispersion. Expressed as Å/mm, it is the difference in the wavelengths of the two lines, in angstrom units, divided by the observed separation of these lines, in millimeters. The lower the number the better the dispersion. The relation between linear dispersion and the angular quantity, $d\theta/d\lambda$, is given by

$$d\lambda/dx = [f(d\theta/d\lambda)]^{-1} \qquad (9)$$

where f is the focal length of the focusing mirror or lens.

For a grating instrument the reciprocal linear dispersion may be found by differentiation of Eq. (2), with the angle of incidence constant, and combining the result with Eq. (9), to give

$$d\lambda/dx = m/(fb\cos r) \simeq m/fb \qquad (10)$$

since $\cos r$ will be virtually constant (r is about 6°). For example, a grating monochromator with a reciprocal linear dispersion of 16 Å/mm, two spectrum lines separated in wavelength by 6 Å, such as 3274-Å Cu and 3280-Å Ag, would be 0.38 mm apart in the plane of the exit slit. Stated another way, using slits 0.100 mm in width, the bandpass would be 1.6 Å—the physical slit width in millimeters multiplied by the reciprocal linear dispersion of the monochromator.

Dispersion for a prism is a function of wavelength. For a medium quartz prism monochromator of focal length 600 mm, typical values of reciprocal linear dispersion would be: 6 Å/mm at 2300 Å, 10.4 Å/mm at 2700 Å, 15.6 Å/mm at 3100 Å, 29 Å/mm at 3700 Å, 54 Å/mm at 4500 Å, and 120 Å/mm at 6000 Å.

B. RESOLUTION

Resolution Rs, also denoted resolving power, is a measure of the ability of a monochromator to discriminate between two very close spectral lines. It is obtained by dividing the wavelength by the separation of the two lines in question. The fundamental resolution of any dispersing device is

$$\lambda/d\lambda = Rs = w(d\theta/d\lambda) \qquad (11)$$

and is limited only by w, the effective aperture width, and the angular dispersion. For example, two distinct lines separated by 0.2 Å at 2500 Å would indicate an achievable resolution of 12,500. However, the resolving power of an actual instrument is generally poorer than the theoretical maximum value because of optical aberrations and other deleterious effects. Also, just what is the criterion for calling two lines resolved? The Rayleigh criterion suggests a 19 % valley between two equally intense lines. Perhaps one would prefer a valley that just attains the baseline between two lines without any stipulation about line intensities. Such a definition would require essentially a bandpass, $\Delta\lambda$, equal to twice the bandwidth of a spectral line at one-half maximum intensity. When narrow entrance and exit slits are used, lines quite close together can be resolved. However, sufficient light must reach the detector to enable the spectral line to be distinguished above the general background signal, thus the light gathering power is also important (q.v.).

For a grating, the effective aperture width is simply the width of an individual ruling b multiplied by the total number of rulings N and by $\cos r$, that is, $bN \cos r$. Since $d\theta/d\lambda = m/b \cos r$,

$$Rs = mN \qquad (12)$$

For example, a grating ruled 600 grooves/mm and 50 mm in width has a resolving power in the first order of 30,000. At the sodium wavelength of 5890 Å, the smallest wavelength interval resolved will be: $\Delta\lambda = 5890/30,000 = 0.2$ Å. It is the product mN that is significant, so that a 50-mm grating of 1200 grooves/mm, where N is 60,000, will have no better resolution in the first order than a 50-mm grating of 600 grooves/mm, where N is only 30,000, when the latter is used in the second order. The advantage of a finer ruling is merely that it permits a higher resolution in the first order.

For perfect image formation, a monochromator must be equipped with curved slits (7). The longer the slit, the greater is the need for a circular slit configuration. Since astigmatism increases with aperture, the curvature becomes more important in fast instruments. The extent of resolution degradation with increasing height of straight slits is shown in Table 1 for a Czerny-Turner 0.5-m grating instrument. Below 2 mm the resolution obtainable with curved and straight slits is equal. Since it is not possible to increase the brightness of a source, there is never an advantage in extending slit height beyond that of the source. Of course, the source may often be extended with an appropriate lens so as to illuminate a long slit.

The resolving power of a prism

$$Rs = t(dn/d\lambda) = (d/f)(dx/d\lambda) \qquad (13)$$

TABLE 1

RESOLUTION DEGRADATION WITH STRAIGHT SLITS

Half-width resolution, Å	Slit height, mm
0.10	2
0.12	4
0.15	8
0.20	18

is limited by the base length of the prism t and the dispersive power of the prism material. The latter is not constant for a prism but increases from long wavelengths to shorter wavelengths. This requires a knowledge of the refractive index of the dispersing material and its rate of change as a function of wavelength, or the linear dispersion as a function of wavelength. A graph supplying this information should be provided with each instrument by the vendor. The effective bandwidth for a specific wavelength is read from the graph, and the curve for the slit width employed.

Over most of the spectrum, the resolving power of an echelle-prism combination is superior to most spectrometers. Table 2 gives, for a few of

TABLE 2

CHARACTERISTICS OF A PRISM-ECHELLE SPECTROMETER[a]

Echelle order	Central wavelength, Å	Dispersion, Å/mm	Resolution, Å	Equivalent slit width,[b] μm
134	1752	2.45	0.08	16
117	2007	2.81	0.09	19
78	3010	4.21	0.14	28
59	3979	5.57	0.19	38
39	6020	8.43	0.28	56
29	8095	11.3	0.38	76

[a] From M. Margoshes, Paper presented at Pittsburgh Conference on Analytical Chemistry and Applied Spectroscopy, 1970.

[b] To achieve stated resolution on a conventional spectrometer with 5-Å/mm dispersion.

the orders, the central wavelength, the dispersion, the resolution, and the equivalent slit width that would have to be used to achieve the stated resolution on a conventional spectrometer with 5-Å/mm reciprocal linear dispersion. The resolution of the echelle alone is given by

$$Rs = (2w \sin i)/\lambda \qquad (14)$$

In use the angle of reflection is usually very nearly the same as the angle of incidence. The free spectral range is about 5 to 10 Å, so that a high dispersion spectrometer or an order-sorter prism must be crossed with the echelle to separate the overlapping orders.

C. Optical Speed

The optical speed is a measure of how fast the optical system can react to or measure a given amount of light. For a photographic plate, the response depends on the illumination; all the energy must be concentrated in the smallest possible image (a silver halide grain) if it is to produce a detectable and sharp blackening of the emulsion. By contrast, a photoreceptor responds to total light flux impinging on its surface; a recognizable signal is one that is distinguishable from background noise. However, if the exit slit is opened until it becomes equal to and then larger than the sensitive area of the photoreceptor, the response of the latter depends once more on the illumination, beyond the point of equality. The photoelectric speed and the resolving power capabilities of a spectrometer are tied together irrevocably and must be considered simultaneously for any meaningful interpretation to result. As the dispersion of an instrument improves, it is possible to achieve a given bandpass with wider slits, thus admitting more light.

The aperture ratio, or f/number, is used to designate the optical speed of a spectrometer. The f/number is given by f/d, the focal length divided by the diameter of the objective. For example, $f/8$ might represent a system involving a 7.5-cm diam collimating mirror of focal length 60 cm. Because the ruled area of a grating (or a prism face) presents a rectangular aperture to the light beam, the effective aperture ratio equates the useful rectangular area to an equivalent circular aperture. The effective aperture area becomes

$$2[(hw \cos i)/\pi]^{1/2}$$

where h is the height, w the width, and i the angle of incidence. For a 25×25 mm grating and a focal length of 250 mm, the effective aperture ratio is $f/8.8$.

In the process of spectral dispersion certain losses are unavoidable. These include reflection losses on the faces of the mirrors and prisms, the loss by absorption in prisms and lenses, and the incomplete utilization of the light beam in the various positions of the dispersing device. For example, in a single glass prism monochromator, 45% of the light of 5500 Å passing through the entrance slit leaves through the exit slit in isolated form. In a double monochromator, the ratio is 20%. For quartz prisms, the corresponding figures are 62 and 38%. In general, the available light decreases with increasing dispersion and higher spectral resolution.

The total flux transmitting power F_T for a monochromator is given by (8):

$$F_T = 4BTsLhw \cos i/\pi f^2 = 4BT(L/f)hw \cos i(d\theta/d\lambda) \qquad (15)$$

where B is the brightness of the source, T is the effective transmission of the optical elements, s is the slit width, L is the slit length, and the other terms have been defined previously. Table 3 gives information for typical spectrometers.

TABLE 3

TOTAL FLUX TRANSMITTING POWERS AND f/NUMBERS OF TYPICAL COMMERCIAL SPECTROMETERS[a]

Instrument	T	h	$w \cos i$	$d\theta/d\lambda$	L/f	Effective f/number	$F_T \times 10^4$
Czerny-Turner $f/8$, 0.25 m, 1180 grooves/mm	0.52	25	24	1.18×10^{-4}	20/250	8.8	16
Ebert, $f/8$, 0.5 m, 1180 grooves/mm, curved slits	0.52	52	50	1.18×10^{-4}	20/500	8.6	64
Czerny-Turner $f/6.8$, 0.75 m, 1180 grooves/mm, straight slits	0.52	102	96	1.18×10^{-4}	18/750	6.8	146
Czerny-Turner $f/6.8$, 0.75 m, 1180 grooves/mm, curved slits	0.52	102	96	1.18×10^{-4}	50/750	6.8	375
Littrow, $f/4.5$ quartz prism, double pass	0.59	70	40	1.48×10^{-4}	1/270	4.5	9

[a] Reprinted from Refs. (8, 9) through the courtesy of Reinhold and Spex Industries, Inc. T = transmission efficiency, h = height of dispersive device, $w \cos i$ = effective width of dispersive device, L/f = slit height/focal length for a resolution of 0.2 Å.

Upon comparing two grating spectrometers, one of which is twice the size of the other, the larger instrument will have twice the flux transmitting power of the smaller, if the slit widths are not altered. However, the grating also plays a role by virtue of its dispersion. The flux transmitting power of an instrument is proportional to the linear dispersion because the slit width is involved. For an equivalent bandwidth and neglecting aberrations or other resolution limiting effects, if the dispersion is increased by a factor of two, the slit width can also be increased by a factor of two, thereby increasing the flux transmitting power by an equivalent amount.

The slit height should be made as long as possible, and the area of the grating should be as large as possible. The focal length should be reduced. Unfortunately, it is not possible to increase the slit height and grating area and simultaneously decrease the focal length beyond certain limits without encountering aberrations which limit the resolving power of the spectrometer.

Both theoretical calculations and experimental data show that, when the slit of a spectrometer is narrowed beyond a certain limit, the illumination drops very rapidly while the resolution is improved only slightly. This point is called the *critical slit width*, and is given by $\lambda f/d$. Thus, the critical slit width on an $f/8$ monochromator for the 3247-Å line of copper would be 3247×10^{-8} cm $\times 8 \equiv 0.0026$ mm, narrower than any practical slit width.

D. Spectral Purity and Stray Light

A monochromator should be as free as possible from scattered light, that is, light outside the narrow spectral waveband determined by the position of the wavelength setting, the dispersion of the prism or grating employed, and the width of the slits. A certain amount of scattered or stray light can never be completely avoided because of ever-present imperfections of all optical parts and surfaces. However, stray light effects are minimized by matte blackening of all internal mechanical parts, by avoiding reflecting edges as far as possible (i.e., mounting mirrors within matte frames), and a careful finish of all optical surfaces. The stray light component is proportional to the height of the slit.

The scattered light situation is vastly improved in a double monochromator whereby the added monochromator provides additional attenuation of the scattered light relative to the primary signal. With single monochromators stray light may be reduced by stray light filters; these are glasses which transmit well within restricted spectral ranges.

Filter-grating monochromators are nearly as efficient as double prism or prism-grating instruments in the attenuation of scattered type light. This is due to the bandpass characteristics of the filters which allow only nearby wavelengths of nearly equal intensity to be transmitted. Here the filter cutoff wavelengths must be considered.

E. WAVELENGTH PRECISION AND ACCURACY

With a single instrument, where standards and samples can be treated alike, repeatability is perhaps the more important. However, wavelength accuracy is desirable, particularly when data from more than one instrument are involved. Wavelength can be checked by measurements on known emission lines from mercury vapor lamps and other discharge lamps. Neon emission lines are readily available since neon is the usual filler gas in hollow-cathode lamps. Table 4 supplies a listing of the mercury and neon lines.

TABLE 4

MERCURY AND NEON EMISSION LINES FOR WAVELENGTH CALIBRATION

Mercury		Neon			
Wavelength, Å	Intensity	Wavelength, Å	Intensity	Wavelength, Å	Intensity
2536.5	1500	2974.7	250	5764.4	700
2967.3	120	3370.0	700	5852.5	2000
3021.5	20	3520.5	1000	5881.9	1000
3125.7	40	4537.8	1000	5944.8	500
3131.6	32	4704.4	1500	5975.6	600
3650.2	280	4708.9	1200	6030.0	1000
3654.8	30	4712.1	1000	6074.3	1000
3663.3	24	4715.3	1500	6143.1	1000
4046.6	180	4752.7	1000	6163.6	1000
4077.8	12	4827.3	1000	6217.3	1000
4358.4	400	4884.9	1000	6266.5	1000
5460.7	320	4957.0	1000	6334.4	1000
5769.6	24	5005.2	500	6383.0	1000
5790.7	28	5037.8	500	6402.3	2000
		5144.9	500	6506.5	1000
		5145.0	500	6599.0	1000
		5330.8	600	8780.6	1000
		5341.1	1000	8783.8	1000
		5400.6	2000	9665.4	1000

F. Optical Alignment

The optimum position of the flame on the optical axis and from the entrance slit of the monochromator depends upon the overall arrangement. In general, it should be at a distance x at which the desired portion of the flame just fills the solid angle embraced by the entrance beam. This is given by the expression

$$x = w(f/d) \qquad (16)$$

where w is the flame width. Often a field lens or entrance mirror before the entrance slit of the monochromator will have been set at the best distance between flame and slit. The lens or mirror serves to image the flame upon whatever aperture within the monochromator, in conjunction with the slits, determines the flux transmission power.

To determine experimentally the optimum flame position, it is only necessary to pass a monochromatic line from a mercury lamp in reverse through the monochromator and observe the light beam at various locations by means of a white card. Lacking a mercury lamp, a continuous source such as a tungsten or deuterium lamp can be used. In an instrument with curved slits, a round circle of light should appear exactly centered above the burner opening. The point of maximum spot brightness establishes the focal point.

IV. Characteristics of Spectrometers and Photometers

A. Different Requirements for Atomic Absorption and Flame Emission

Instrument requirements are different for spectrometers to be used in atomic absorption as opposed to those for use in flame emission. Generally speaking, requirements are more stringent for the flame emission spectrometer.

1. Bandpass

In atomic absorption the specifications for bandpass are relatively simple. The monochromator must be able to pass the resonance line, and yet screen out any nonresonant lines including those of the filler gas in the hollow-cathode lamp in order to achieve a straight calibration curve.

For critical separations, such as iron, cobalt, nickel, and the rare earths, one needs about a 2-Å bandpass. A larger bandpass will cause the absorbance curve to flatten and bend, whereas a smaller bandpass will produce a reduction in the available light and therefore a degradation of the signal-to-noise ratio. The effect of increasing the bandpass on the absorption of the arsenic line at 1937 Å is shown in Table 5. Although arsenic is reported

TABLE 5

ROLE OF HALF-BANDWIDTH ON THE
ABSORBANCE OF ARSENIC AT 1937 Å[a]

Half-Bandwidth, Å	Absorbance[b]
0.2	0.229
0.4	0.222
0.6	0.212
0.8	0.202
1.2	0.169
1.6	0.158
2.4	0.158
4.0	0.143

[a] Reprinted from Ref. (10).
[b] Arsenic concentration, 100 μg/ml, using argon (entrained air)–hydrogen flame with total consumption burner.

to be relatively free of spectral interferences, the data would indicate that some nonabsorbing lines are in the region of the arsenic resonance line. Should the spectral bandwidth (twice the bandpass) exceed 7 Å, the neon line at 1930 Å would interfere; there is also an ionic neon line at 1939 Å.

To be able to adjust the monochromator to various bandpasses and, in atomic absorption to adjust the intensity of the source beam, variable slits are necessary. As the slits are widened, commensurate with adequate resolution, more light is passed through the monochromator, and this can be translated into an improvement in precision and detection limit in emission flame spectrometry.

In flame emission, spectrometers with high resolving power are needed to handle spectra from line-rich elements and closely spaced multiplets, whereas molecular emission bands show up better with instruments of low dispersion. A test of an instrument's resolving power concerns the separation from each other, of the manganese triplet at 4033 Å, the potassium doublet at 4044 Å, and the lead line at 4058 Å. Only grating spectrometers possess the required resolution. On the other hand, when working with the fluctuation bands of BO_2, a medium quartz spectrometer is preferred since it will not resolve the individual vibrational components of each band. A similar situation prevails when using the oxide band systems of the rare earths.

2. Recording Systems

For work in flame emission the spectrometer should be able to scan through the wavelength region of interest so as to record the background emission on either side of the emission line or bandhead. Preferably, several scanning rates should be possible. Often atomic absorption signals are recorded on a time scale in order to estimate the "average" absorption signal of analyte, standard, and background blank, separately.

In all measurements involving electronic recording, there is a maximum possible instrumental scanning speed allowable for any set value of the slit width, amplifier parameters, and recorder pen traverse time (11). By definition, the time constant of an electronic detection system is the time required for the measuring device to reach 63%, i.e., $100 - 100/e$, of its maximum scale value when a step-function signal is received at the detector. Roughly four time constants are required for one full peak height, or 98%. This is often referred to as the response time. Assuming that the recorder pen traverse speed is faster than the time constant of the electronics producing the signal being recorded, the maximum scanning speed, in Å/sec, to produce a faithful record of a spectrum is equal to the bandpass, in angstroms, divided by the response time, in sec:

$$\text{Maximum scan speed} = \text{bandpass/response time} \qquad (17)$$

$$= \text{bandpass/}(4 \times \text{time constant}) \qquad (18)$$

Increasing the amplification involves an increase in noise, which is proportional to the square root of the amplification. This requires a longer response time to dampen out fluctuations. Thus the response time should be increased in proportion to the square root of the change in amplification. For example, when one switches from a $1 \times$ to a $10 \times$ scale (10 times amplification increase), the response time would be increased by about three and, correspondingly, the scanning speed should be reduced.

B. Filter Photometers Compared with Monochromators

In a filter photometer (Chapter 9, Fig. 11) an interference filter replaces the monochromator. This change restricts the system to one wavelength per filter; however, the relatively low cost of interference filters permits a quantity of filters to be purchased for the price of an inexpensive monochromator. When using filters, the aperture to the detector can easily be opened to a centimeter in diameter with condenser lenses, resulting in an effective detector aperture of 0.785 cm² as compared to the 10^{-6} mm² aperture of a monochromator. Light from the desired region of the flame is collimated by a lens of heat-resistant glass, so that the condensing lens system sends the light through the interference filter as an approximately parallel beam; this is especially important to achieve the peak wavelength. The photocurrent is approximately maximal when, with sufficiently large photosensitive area, the flame is imaged on the detector at a magnification of unity. For flame emission a mirror should be placed behind the flame to increase the light intensity about twofold. The detector receives a signal many magnitudes greater with the filter, allowing the use of vacuum photodiodes or semiconductor devices rather than multiplier phototubes. Not only do these detectors have better noise characteristics, but they are much less expensive. In the final analysis, the filter system achieves a considerable improvement in the signal-to-noise ratio and allows a much greater precision to be achieved. The stability of filters offers definite advantages for routine or repetitive analysis. Since the peak wavelength and bandpass are fixed, pushbutton selection is possible. There is no danger of incorrect wavelength setting or drift from the peak wavelength during operation, necessitating readjustment or recalibration. Admittedly there is a limitation in applications where the investigator may wish to experiment with a number of different resonant lines, emission lines, or bandwidths. Lamp-filter combinations are not presently available for all the elements determinable by atomic absorption. Filters are not available with sufficiently well-defined transmission bands in the ultraviolet below 3600 Å.

A monochromator can accept only a small part of the radiation from the flame because of the small entrance slit. Hence, the detection limits in flame emission are often fixed less by the optical resolution than by the ratio of signal to noise. By comparison with filter photometers, any wavelength can be selected on the monochromator and the wavelength range can be extended into the ultraviolet with quartz or fused silica optics. Hence monochromators permit determination of many additional elements and offer a larger choice of analytical lines. More skill is required to operate a spectrometer; choice of appropriate slit widths, peaking the spectral line (except in scanning instruments), and selection of flame region to be viewed, all require experience.

C. SINGLE-BEAM, DOUBLE-BEAM, AND DUAL-WAVELENGTH SYSTEMS

In single-beam operation for atomic absorption the output of the amplifier is arranged to give full-scale deflection with no sample in the flame. By contrast, for flame emission the detector-amplifier circuit is adjusted to give a zero reading and suppressing, if necessary, any flame background emission. There is only one set of optics. In a flame photometer, light emitted from the center of the flame just above the inner cone is focused by a lens of heat-resistant glass and passes through interchangeable filters onto a single photoreceptor, or light from the proper location in the flame gases passes into a monochromator and radiation leaving the exit slit is focused onto the detector.

In double-beam flame emission photometers, a second light path is provided for the light emitted by the internal standard element. The internal standard and analytical wavelengths are isolated by means of optical filters in a dual optical system, and each beam is focused on a separate photodetector. The signal of one detector opposes that of the other through a galvanometer. Light usually leaves the flame through two apertures on different sides of the flame and then through the respective optical paths. However, the flickering of the flame is not necessarily directionally isotropic. To eliminate flicker completely in a double-beam system, the photoreceivers must be arranged to view the flame from the same direction by use of a semitransparent (half-silvered) mirror.

In atomic absorption a double-beam system cannot overcome instability and noise from the burner, since the burner is only in one of the two beams (Fig. 14). The chief advantage of a double-beam unit is in overcoming the effects of source drift and change in detector sensitivity. Hollow-cathode lamps require from 5 to 30 min to stabilize after being turned on. A double-beam system may be used during this period, thus making rapid lamp change quite convenient. The improved baseline stability makes it possible to see small departures from it, thereby improving detection limits. More importantly, the stability when the flame is almost completely transparent makes it possible to use electrical scale expansion techniques to improve the detection limit. At higher concentrations, analytical precision may be improved by using differential spectrometric techniques

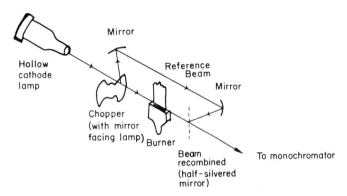

Fig. 14. Double-beam optical system for atomic absorption.

whereby the zero signal is suppressed and a small signal difference is expanded to full scale. Among the liabilities of the double-beam system is the lengthened optical path and the increased instrumental complexity. Diversion of roughly half of the source light lowers the flux transmission through the flame correspondingly. A substantial percentage of the analyses for which atomic absorption is being used and contemplated can be adequately performed by a single-beam instrument.

In a dual-wavelength system, as shown in Fig. 15, two interference filters (or a dual monochromator) are alternately interposed in the optical path between the flame and the detector. The resonant line filter isolates the

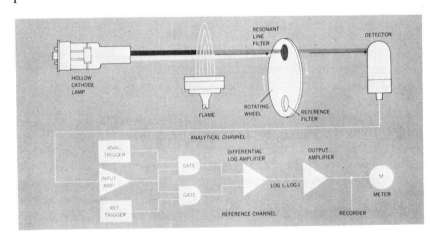

Fig. 15. Dual-wavelength system.

proper analytical line for each element; a separate filter for each element is usually required. The second filter passes selected neon or argon reference lines emitted by the filler gas of the hollow-cathode source. Thus the detector alternately measures the intensities of the resonant line and the reference line. This dual-wavelength system compensates for minor lamp fluctuations and variations in the detector-amplifier electronics. Unlike the double-beam approach, the reference radiation follows exactly the same path as the resonant radiation, which minimizes the effects of light scattering and dispersion within the flame (see Chapter 10 of Volume 1).

D. MULTICHANNEL INSTRUMENTS

On multichannel instruments a number of exit slits in the focal plane of a monochromator are used to isolate several individual wavelengths. The optical intensities at each selected wavelength are converted by a photomultiplier into electrical signals. One circuit is commonly reserved for an internal standard element.

When both atomic absorption and atomic emission techniques are employed simultaneously, the atomic absorption source is modulated at one frequency, and the light emitted by excited atoms in the flame is modulated at a different frequency. To do this, separate choppers are placed before and after the flame, respectively. The block diagram shown in Fig. 16 illustrates how each of these functions is performed. The spectral information is encoded in such a way that the data pertaining to each element may be electronically separated and displayed. To utilize the information inherent in the emission spectra, this unit incorporates an echelle grating. With an echelle spectrometer all wavelengths are measured on the grating blaze. The luminosity therefore remains high over the entire spectral range which makes the instrument very well suited for weak or short-duration light sources.

In a unique multichannel spectrometer, Mavrodineanu and Hughes (12) accomplished the simultaneous atomic absorption and flame emission analysis of three elements by each mode. As shown in Fig. 17, the characteristic radiations from individual hollow-cathode tubes were combined into a single, collimated polychromatic beam that is passed through the flame, then resolved into its components, and each component brought onto a detector. The source radiations are combined by locating each source behind a slit that is so positioned with respect to a diffraction (or dispersion) device to satisfy the diffraction (or dispersion) equations under

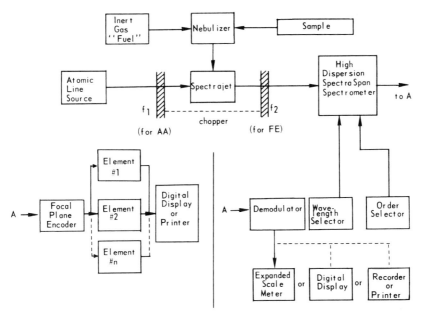

Fig. 16. Multichannel operation utilizing both atomic emission and atomic absorption. (Courtesy of Spectrametrics, Inc.)

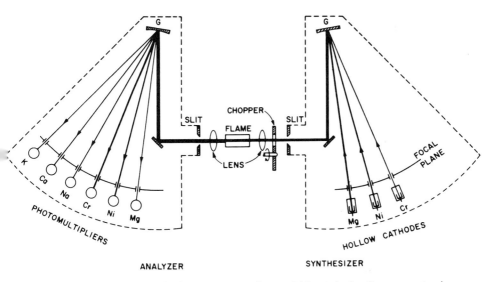

Fig. 17. Schematic optical arrangement for multichannel simultaneous atomic absorption and atomic emission spectrometry. [From Ref. (*12*) by courtesy Applied Optics.)

conditions which bring each beam along a common path to a single exit slit. This is the synthesizer unit. The analyzer unit is a conventional spectrometer with additional slits fitted in the focal plane at positions corresponding to the atomic radiations excited in the flame and thus permit the determination of elements in the emission mode, as well as by atomic absorption.

Multichannel spectrometers are severely limited in the number and locations of the wavelengths at which intensities can be measured at one time, and it is extremely difficult and time consuming to change the line array. Only a small fraction of the total information in a spectrum can actually be measured and made available for use. However, for routine analytical applications multichannel instruments are ideal.

E. INTERFEROMETRIC SPECTROMETER

A modification of the Michelson interferometer provides up to 1000 to 1 rejection of background radiation while retaining the high optical efficiency

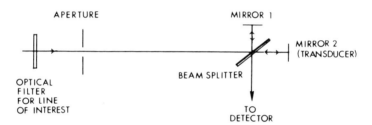

Fig. 18. Optical diagram of interferometric line discriminator. (Courtesy of Block Engineering, Inc.)

of interferometric spectrometers. The basic configuration is shown in Fig. 18. Light falling on the beam splitter is divided into two beams for mirrors 1 and 2. These mirrors are positioned so that the beams travel the same optical path distance in reaching the detector. Mirror 2, however, is mounted on a transducer so that its position can be varied, thus changing the optical path distance of the second beam. If light of wavelength λ enters the aperture, and mirror 2 is moved back a distance $\lambda/4$, then beam 2 will travel a distance $\lambda/2$ greater than beam 1. Consequently, beam 2 arrives at the detector 180° out of phase with beam 1, and the beams cancel. As mirror 2 is moved back and forth, an alternate lightening and darkening is seen at the detector, the frequency of which is proportional to the wavelength of the incident radiation. In effect, the interferometer optics are designed to produce a "comb" of transmission channels which are moved back and forth along the wavelength axis. The transmission of the energy in a particular spectral line goes alternately from 0 to 100%. This modulation of the spectral line is seen as an ac signal by the detector. However, since the background radiation contains energy at all wavelengths, it will not be modulated by the transmission channels, and dc background noise will not be seen by the detector. The output reading will be directly proportional to the intensity of the spectral line.

The entrance aperture of the interferometer is roughly 5000 times larger in area than the entrance slit of a conventional spectrometer. As a result a far greater amount of energy enters the instrument. A further advantage is the fact that the field of view of the optical system is 15°, eliminating the need for exact alignment optically.

REFERENCES

1. H. H. Willard, L. L. Merritt, and J. A. Dean, *Instrumental Methods of Analysis*, 4th ed., Van Nostrand, Princeton, New Jersey, 1965, p. 40.
2. H. Ebert, *Wied. Ann.*, **38**, 489 (1889).
3. M. Czerny and A. F. Turner, *Z. Physik*, **61**, 792 (1930).
4. W. G. Fastie, *J. Opt. Soc. Am.*, **42**, 641 (1952).
5. G. R. Harrison, *J. Opt. Soc. Am.*, **39**, 522 (1949).
6. *The Spex Speaker*, 8 (No. 4) (December 1963).
7. W. G. Fastie, H. M. Crosswhite, and P. Gloersen, *J. Opt. Soc. Am.*, **48**, 106 (1958).
8. R. F. Jarrell in *The Encyclopedia of Spectroscopy* (G. L. Clark, ed.), Reinhold, New York, 1960, pp. 247–250.
9. *The Spex Speaker*, 9 (No. 3)(October 1964).
10. O. Menis, ed., NBS Tech. Note 504, 1969.
11. J. R. Adkins, *Ramalogs*, **1** (No. 1)(August 1968).
12. R. Mavrodineanu and R. C. Hughes, *Appl. Opt.*, **7**, 1281 (1968).

6 Electronics

C. Veillon

DEPARTMENT OF CHEMISTRY
UNIVERSITY OF HOUSTON
HOUSTON, TEXAS

I. Introduction

The basic electronics most often used in flame spectrometry consist of
a source power supply, a photosensitive detector of some sort (plus any

149

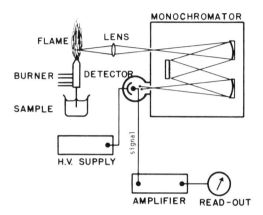

Fig. 1. Basic instrumental configuration for emission flame spectrometry.

necessary power supplies), an amplifier, and some type of readout device. Although many variations of each of these exist, these basic components are the ones almost always used. The notable departures from this group are the omission of the external source and its power supply in flame emission, and a photocell-galvanometer arrangement is used in some filter flame photometers (emission), omitting the amplifier and high voltage power supply.

The many variations of these basic components depend on the method (emission, absorption, or fluorescence), the optics (single-beam, double-beam, dual double-beam, or combinations), and the sensitivity and

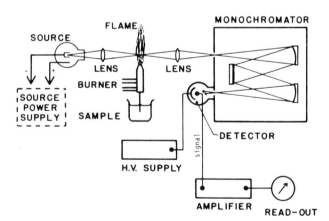

Fig. 2. Basic instrumental configuration for atomic absorption spectrometry.

detection limit (i.e., signal-to-noise ratio, S/N) required. However, the similarities are illustrated in Figs. 1, 2, and 3. The only essential difference, in these illustrations of the basic instrumental configurations for emission, absorption, and fluorescence, is the optical arrangement. The external source needed for absorption and fluorescence will, of course, require an additional power supply, depending on the particular type of source used (see Chapter 2).

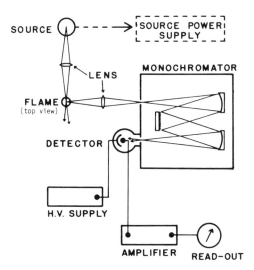

Fig. 3. Basic instrumental configuration for atomic fluorescence spectrometry.

II. Power Supplies

Other than the power supply for the source for absorption and fluorescence (Chapter 2), the only other power supply usually required in flame spectrometry is a well-regulated, stable source of high voltage direct current for the detector. The capabilities of this power supply must, of course, be adequate to meet the demands of the particular detector used. More will be said on the detectors themselves in the following section. Since the most commonly used detectors requiring an external power supply are the vacuum phototube and the multiplier phototube, the discussion will emphasize power supplies for these, especially the latter. These devices require voltages ranging from under -500 V to over -2000 V, most being between -500 and -1500 V. The voltages are *negative* (with respect to ground), and the positive terminal is usually at ground potential.

In some of the earlier emission flame spectrometers, or "flame photometers" as they are still often called, batteries (dry cell) connected in series were used to provide the phototube power. This cumbersome arrangement, although functional, has all but been eliminated by electronic power supplies, whether they be vacuum tube, solid state, or hybrid types. Inexpensive compact power supplies are available, with outputs of -2000 V or more, at several milliamperes, regulated to within 0.001% against line voltage and load current changes, stable to within 0.01% over periods of hours, and having ac ripple components less than 1 mV. Specifications such as these represent a very good power supply and are to be considered minimum requirements for high precision work. For most routine applications, power supplies with poorer specifications can be and are used, but the cost saving is usually small.

III. Photosensitive Detectors

A photosensitive detector is simply a "light transducer" which converts radiation into some measurable quantity (usually electrical) compatible with the readout system employed. Because flame spectrometry has been limited almost entirely to the visible and ultraviolet regions of the spectrum, many of the available photosensitive detectors are not used, due to their poor response in these regions. For example, the bolometers and Golay cells used in the infrared region are of little use in flame spectrometry.

Of the detectors that have been used—thermopiles, photovoltaic cells, photoconductive cells, photographic emulsions, phototubes, and multiplier phototubes—the last is by far the most popular. However, a brief description of the others will be given.

A. THERMOPILES

These are thermocouples, or banks of thermocouples, coated with a black absorber. Output is a voltage (emf) of low impedance. Although used mostly in the infrared region, they can be used in the visible and ultraviolet. One important characteristic of these detectors is their virtually complete independence of wavelength sensitivity; that is, for a given incident radiant power the output is independent of the wavelength of the radiation. Thus, this is essentially a "flat-response" detector. This characteristic is highly desirable for certain theoretical aspects of flame spectrometry, such as calibrating the spectral output of sources in absolute units. These data can then be used to calibrate detectors, determine transmission characteristics of monochromators, and so on. There are some problems associated with this detector, not the least of which is their low output impedance, making it difficult to amplify the signal. This problem can be overcome to some extent by modulating (chopping) the light beam and using an impedance-matching transformer and an ac amplifier, usually at a relatively low frequency because of the response time of these detectors. Except for a few specialized problems, thermopiles are not used in flame spectrometry.

B. PHOTOGRAPHIC EMULSIONS

Since the very early days of spectroscopy, photographic films and plates as photodetectors have been used. They are still widely used today, particularly in emission spectrographic analyses, and have several unique advantages: a large spectral region is recorded simultaneously, entirely and permanently. They are "integrating" detectors, i.e., they can measure the *total* light flux at the various wavelengths for successive exposures; they are (or can be) sensitive over very wide wavelength regions, from the near-infrared down to the far-ultraviolet, and even beyond into the X-ray region in appropriate instruments. Sensitivity is good, and in many applications can be exceptional because of their ability to integrate the radiation flux.

Disadvantages of photographic emulsions as photodetectors include: relatively poor reproducibility, and hence, accuracy; variation of sensitivity with wavelength; variation between emulsion "batches," and even between plates or films of the same emulsion batch; a time-consuming development process to bring out the latent information; another time-consuming measurement process to "read" the information available. The latter process involves the use of a microdensitometer to measure line blackening and/or wavelength. Though widely used in spectrographic analysis, and in the past for emission flame spectroscopy, photographic detection is rarely used anymore, especially for quantitative work.

C. PHOTOVOLTAIC CELLS

These cells are commonly known as barrier-layer cells or simply "photocells." They consist of a semiconductor, usually selenium, deposited on a conductor, like iron, and on exposure to light, electrons flow across the junction (or "barrier") from the semiconductor to the conductor, generating an emf output. Silicon is also used in some cells, as well as other semiconductor-conductor combinations. Because of the low impedance of the cell, amplification of the signal is difficult and therefore these detectors have been and are used mostly in filter photometers, where the large optical aperture and high light flux generate sufficiently large signals for use without amplification. Sensitivity of the cells is moderate, high sensitivity depending on the optical aperture (area of sensitive surface) and external resistance. Wavelength sensitivity approximates that of the human eye. High flux levels tend to "fatigue" these cells, the temperature coefficient is appreciable, and the response time relatively slow ($\sim$10 msec or more). Linearity of response is good, provided the light flux and external resistance are kept low.

D. PHOTOCONDUCTIVE CELLS

As the name implies, these devices undergo a conductivity change on exposure to light. Essentially they are photosensitive resistors, having high dark resistance which decreases on exposure. If an external dc voltage is applied across the cell, the current flowing through the cell will increase as the light flux increases. Lead sulfide and lead selenide cells are often used in the red-visible and near-infrared region, but some of the newer cadmium sulfide and cadmium selenide cells are superior at shorter wavelengths. For example, one common type has an S-15 response curve, covering a wavelength range from about 3000 Å to over 8000 Å, with maximum response between 5000 and 6000 Å. This wavelength coverage should be ideal for emission flame spectrometers, since the energy in most chemical flames restricts atomic emission to this approximate wavelength range.

Photoconductivity cells are not often used in simple emission flame spectrometers, but offer many advantages for this purpose. Their small size, high sensitivity, high impedance, rapid response, and spectral sensitivity should make these devices superior (to photovoltaic cells) detectors for simple filter flame photometers. Used in conjunction with the highly selective interference filters available today, it should be possible to design a compact, relatively simple, portable, multielement emission flame spectrometer.

E. PHOTOTUBES

By far the most commonly used detectors in flame spectrometry, phototubes may be divided into two groups: the simple vacuum phototube, utilizing the ejected photoelectrons from a suitable cathode, and the multiplier phototubes, utilizing, in addition to the photoelectrons, electrons obtained from secondary emission from one or more dynodes or "stages."

A simple phototube is shown schematically in Fig. 4. Entering photons strike the cathode and eject photoelectrons, which are collected at the

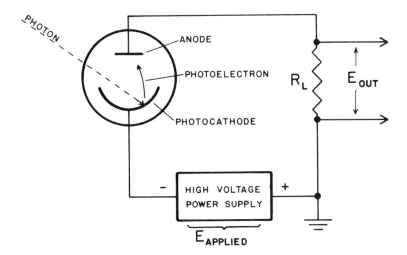

Fig. 4. Schematic of a simple vacuum phototube.

more-positive anode. Ejection of electrons at the cathode by photons is the well-known photoelectric effect. Cathodes are usually coated with various materials (e.g., the alkalis) selected so that the work function is low, thermionic emission is low, and/or the desired spectral response is obtained. Above a certain voltage (that at which essentially all ejected electrons are collected), the response (current) is proportional to the light flux, and essentially independent of the applied voltage at a given flux. This is illustrated for a typical phototube in Fig. 5.

These detectors are simple, rugged, relatively sensitive, stable, but not used much anymore. They have been largely replaced (except in simple instruments) by the multiplier phototube.

The multiplier phototube, or "photomultiplier" as it is commonly called, is by far the most widely used detector in flame spectrometry. These detectors are simple phototubes with built-in dynodes for current multiplication by secondary electron emission. This can be shown schematically for a 3-stage photomultiplier as in Fig. 6. A typical electrical schematic is also shown in Fig. 6. Note that the negative terminal of the high voltage

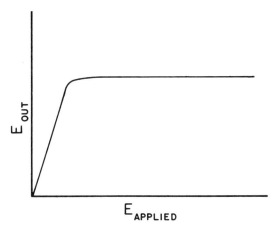

Fig. 5. Response vs voltage at constant light flux for a typical phototube.

supply is connected to the photocathode, and the positive terminal to the ground. Note also that the anode is also connected to ground, through the load resistance, R_L. Assuming that the dynode resistors R are all of equal resistance, and assuming -400 V applied to the cathode, then the current flow through the dynode circuit will produce a 100-V drop in voltage between each dynode, i.e., each succeeding dynode is 100 V more positive than the one preceding it. Also, the anode, being essentially at ground potential, is 100 V more positive than the last dynode.

A single photon strikes the cathode and ejects a single photoelectron. This electron is accelerated through a drop of 100 V to the first dynode which strikes with sufficient force to eject one or more *secondary* electrons. Each of these are now accelerated through a drop of 100 V to the second dynode where this process is repeated. If two are ejected at the first dynode, four at the second, and eight at the third, and these collected at the anode, a "multiplication factor" of 8 is obtained. The resultant current flows to ground through the load resistor R_L, producing a voltage drop across R_L proportional to the current and, therefore, proportional to the light flux. R_L may be installed external to the amplifier as shown, or may be part of the amplifier input circuitry. Needless to say, the greater the input impedance (resistance) of the amplifier, the greater the value of R_L which can be used, and, therefore, the greater the voltage drop for a given current. This being but one possible dynode circuit, other useful circuits are given elsewhere, such as in the book by Lion (*1*).

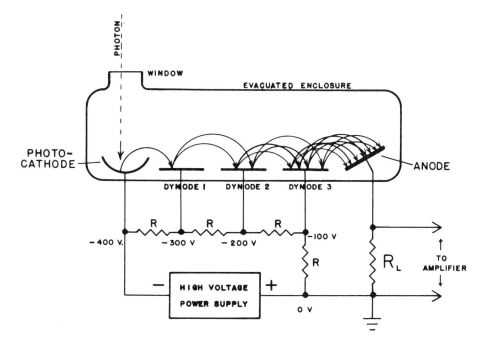

Fig. 6. Schematic of a photomultiplier and attendant power supply (400 V).

In practice, several secondary electrons are given off at each dynode for each incident electron, and most commonly used photomultipliers have between 8 and 15 stages (dynodes). If the number of secondary electrons emitted per incident electron is X, and there are n dynodes, then the overall current multiplication (called "gain") is X^n. For example, if $X = 4$ and there are 10 dynodes, the gain is about 10^6. Gain of the photomultiplier can be controlled over wide limits by changing the overall applied voltage, and therefore the voltage per dynode (see Fig. 6).

One can readily see the extreme sensitivity of photomultipliers, as compared to the previous detectors. Extremely low light levels can be measured; under proper conditions individual photons can be counted. Many photomultipliers have linear response (current vs light flux, at constant voltage) over 5 or 6 orders of magnitude (e.g., 10^{-9} to 10^{-3} A). Response time is virtually instantaneous, being on the order of nanoseconds. Gain stability is excellent, both with respect to light flux level and time. Spectral response varies, depending on the cathode and window materials, with most tubes operating in the region of 2000–6000 Å.

Some of the more popular photomultiplier tubes used in flame spectrometry include the following: Radio Corporation of America (RCA), 931A, 1P21, and 1P28; EMI Electronics, Ltd. (EMI), 6256S and 9558Q; and Hamamatsu TV Co. Ltd. (HTV), R106, R136, and R212.

The RCA types 931A and 1P21 are virtually the same, but the latter has higher gain and lower noise. Response is S-4 (3000–6500 Å). The 1P28 is probably the most popular in flame spectrometry, but is being replaced by the HTV R106, which has a better short wavelength response and lower noise. The 1P28 has S-5 response (2000–6500 Å).

The HTV type R106 is essentially a 1P28 with a fused silica window (envelope). Spectral response (S-19) is consequently improved (1650–6500 Å). The R136 does not yet have an S-number assigned, but is a wide-range tube, having good response from less than 2000 Å to above 7500 Å. The R212 has S-5 response and is virtually identical to the 1P28, except for lower specified anode dark current.

The EMI tubes are of the "end-window" types, as opposed to the previous types, which are side-window types. Also, the previous tubes are all 9-stage photomultipliers, while the 9558 and 6256 have 11 and 13 dynodes, respectively. The EMI photomultipliers are extremely well made and have excellent characteristics. The types mentioned are just two of many EMI tubes, but these two seem to be the most popular in flame spectrometric applications. The 6256S has an "S" (Q) response (1650–6500 Å). Operated under conditions where its gain is about an order of magnitude greater than the R106, the noise is nearly two orders of magnitude *lower*. The 9558Q has an S-20 (Q) response (1650–8500 Å), making it one of the widest-range photomultipliers available. With this single tube, radiation from the far-uv to the near-ir can be detected. Operated at the same gain as an R106, the noise is comparable, being about half that of the R106. This is quite amazing, when one considers the noise characteristics of most red-sensitive photomultipliers.

Noise in photomultipliers used in flame spectrometry results primarily from three effects: "dark current," the "shot" effect, and the "Johnson" effect. Dark current (current passing through the tube even in the absence of light) is due primarily to thermionic emission (cathode and dynodes), and to leakage of current over the glass envelope. The thermionically emitted current is proportional to $AT^2e^{-\theta/T}$ where T is the absolute temperature, A is the cathode area, and θ is the work function of the emitting surface. Thus, one can readily see that the thermionic component of the dark current can be reduced by cooling the surface responsible (usually the cathode) and by decreasing the cathode area. The thermionic emission increases rapidly as θ decreases, i.e., as the photoelectric threshold wavelength is shifted toward the red (longer wavelengths). This is one reason why red-sensitive photomultipliers are usually much noisier than those with shorter wavelength response. By response here is meant the maximum (longest) wavelength to which the tube will respond. Electrical leakage over the glass envelope (at a given voltage) can be reduced to a minimum by carefully cleaning the envelope and maintaining it in a dry environment. Coating the surface with a nonconductive, water-repellant compound will also prevent moisture adsorption and reduce leakage.

Since dark current (i.e., thermionic and leakage current) is essentially dc it can be effectively eliminated by modulating the source radiation, thus producing an ac photoelectric current. This will be described in greater detail in the later discussion of amplifiers.

Shot noise, which is due to the statistical (random) fluctuations in the electron current through the tube, produces an ac current superimposed on the tube current. The frequency of this noise is uniformly distributed over the frequency spectrum (so-called "white noise") from 0 to 10^8 Hz (2). However, the measurable part of this ac current is limited by the frequency response band width (Δf) of the associated electronics and readout device. It is proportional to $(\Delta f)^{1/2}$ at a given average emission current in the tube, and hence can be reduced by decreasing the frequency response band width of the associated electronics. This, too, will be discussed in the section on amplifiers. Shot noise can also be expressed as a voltage (current through a load resistor R_L, as in Fig. 6), in which case the noise voltage is directly proportional to the value of R_L, as well as to $(\Delta f)^{1/2}$.

Johnson noise is due primarily to the thermal agitation of electrons in a resistor, usually the load resistor (Fig. 6). This ac noise, expressed as a voltage across R_L, is proportional to $(TR_L \Delta f)^{1/2}$, where T is the absolute temperature.

In flame spectrometry, be it emission, absorption, or fluorescence, the magnitude of the noise is not as important as the signal-to-noise ratio (S/N). Some consideration of this will be given later, but those interested in the S/N ratio of photomultipliers *per se* should consult the paper by Engstrom (*3*), as well as the texts by Rodda (*4*) and Lion (*1*). The two latter ones are also informative on practically all aspects of photomultipliers, as well as other detectors.

One problem with photomultipliers, frequently encountered but not always recognized, is that of "fatigue." If a photomultiplier is exposed to too high a light flux, reversible (and sometimes irreversible) reduction of the sensitivity can occur. This fatigue effect, the decrease in the tubes' sensitivity, is caused primarily when the high rate of emission of photoelectrons is faster than the electrons can be replenished from internal layers, leaving unneutralized positive sites on the cathode surface. In addition to the decrease in sensitivity, a shift in the spectral response of the tube may occur, usually shifting the threshhold to shorter wavelengths. Often, the photocathode will recover after a period of darkness (several minutes to several hours at room temperature). For most tubes, a continuous *anode* current of 1 mA or less will require about one day of darkness for recovery. At higher currents, the damage may be irreversible (*3, 5*).

For most tubes, if long-term stability is desired, the anode (output) current should be limited to 10^{-4} A or less. For short durations (few seconds), average currents up to 10^{-3} A can be tolerated. For pulsed operation, much higher currents can be tolerated, but this mode of operation is not often employed in flame spectrometry, despite its advantages. Apparently no one has yet had the time and/or inclination to develop this detection technique for flame spectrometry.

IV. Amplifiers

Except in the case of some simple filter-type emission instruments employing photovoltaic cells, some means of amplification of the signal from the detector is usually employed. Amplifiers may be divided into two general types: ac and dc, and some will amplify both ac and dc signals. Since a large book could be written on this subject alone, the discussion will be restricted to those basic types of amplifiers commonly employed in flame spectrometers and to the more important aspects (from a practical useage point of view) of amplifiers.

A. PRINCIPLES OF AMPLIFICATION

Assuming a detector circuit like the one shown in Fig. 6, the anode current (ac or dc) will develop a voltage (ac or dc) across the load resistor R_L. An ac signal could be produced, for example, by employing a mechanical "chopper" to modulate the light generating the anode current. Many configurations of these choppers are possible, but most are a rotating disk with opaque and transparent sections (e.g., holes) which alternately block and pass the radiation, producing an ac signal. This ac signal could then be amplified by a circuit like that shown in Fig. 7, e.g., a triode amplifier.

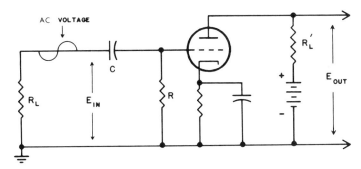

Fig. 7. Triode amplifier.

In Fig. 7, R_L is the source load resistor (Fig. 6), across which the ac voltage E_{in} appears. R is the input resistance of the amplifier, and provides a path to ground for those electrons (in the triode) collected by the grid. The coupling capacitor C connects the ac signal to the grid but does not allow any dc signals to pass. The small ac signal then modulates the much larger triode current resulting in a much larger ac voltage (E_{out}) across the triode load resistor R'_L. If the output of this amplifier is connected to the input of another similar amplifier, a further increase in amplification can be obtained. Further detailed information about the operation of amplifiers *per se* may be found in texts on the subject of electronics, such as that of Malmstadt et al. (6).

A dc amplifier could be obtained from the same basic circuit shown in Fig. 7 by simply omitting the coupling capacitor C, resulting in a *direct-coupled* amplifier. One problem associated with this type of direct-coupled dc amplifier is drift. This type of circuit is more or less limited to only one stage of amplification, since any changes in the first stage (due to temperature changes, aging of components, etc.) will be amplified by successive stages. To eliminate the problem of drift, two systems are usually available: (1) modulate the light producing the signal and use an ac amplifier; (2) modulate the incoming dc signal (converting it to an ac signal) and using an ac amplifier. By coupling the ac amplifier stages with capacitors, one can then block the passage of dc and eliminate drift. This is the familiar "chopper-stabilized" amplifier. In atomic absorption and atomic fluorescence, system (1) is usually employed—chopping the source radiation and using an ac amplifier—because the essentially dc signal resulting from the un-chopped flame emission is therefore not amplified (see Figs. 1–3). In flame emission spectrometry, both systems (1) and (2) are used, with more or less comparable results. However, there are some advantages to system (1) for emission spectrometry, when one considers the effect of noise (all sources), as we shall see in the following sections. Normally, the output of the last amplifier stage is rectified (converted to dc) and filtered (to remove ac components, or "ripple") before appearing at the amplifiers' output terminals. This allows one to use somewhat less complex readout devices, such as potentiometric strip-chart recorders, meters (crude galvanometers), or analog-to-digital converters (e.g., digital voltmeter). Hence, one obtains a dc output proportional to the input voltage or current, be it ac or initially dc which is subsequently modulated.

B. SIGNAL INPUT

Some other aspects of amplifiers which must be considered are signal input, output, frequency response, and noise. When connecting the output of a photodetector to the input of an amplifier, one must be careful that the relative impedances (resistances) are compatible. When employing a direct current, a frequency of 0 Hz, one may consider the impedance and resistance to be the same. At high frequencies, the total impedance is not due to the purely resistive components, but also to any capacitance or inductance present in the input circuit [see Ref. (6), Supplement 3]. However, at the low frequencies usually employed in flame spectrometry, resistance and impedance may be considered to be essentially the same. In general, it is desirable that the *input* impedance of a device (e.g., an amplifier) have a much higher impedance than the *output* impedance of the device to which it is connected. This assures maximum transfer (e.g., of voltage) from one to the other. Let us take as an example a photomultiplier as in Fig. 6. The anode current produces a corresponding voltage across the load resistor R_L, which is fed into the amplifier input. This load resistor may be part of the photomultiplier circuit, as shown, or simply the input impedance of the amplifier. The photomultiplier tube itself (without R_L) has an extremely high output resistance, being the resistance between the anode and last dynode (as in Fig. 6) across the glass envelope. With this essentially infinite output resistance, R_L can be very large, which is an important advantage. The larger R_L is, the greater the voltage drop across R_L for a given anode current. To be able to use a very high value of R_L (Fig. 6), the input impedance (resistance) of the amplifier must be very large, much larger than R_L, otherwise voltage division will occur. For example, if R_L is 10^7 ohms and the input impedance of the amplifier (connected across R_L) is only 10^6 ohms, then the resulting parallel resistance has a net value of only R_n, given by

$$\frac{1}{R_n} = \frac{1}{10^6} + \frac{1}{10^7} \tag{1}$$

or $R_n = 9.1 \times 10^5$ ohms. Therefore, the net load resistance is essentially that of the amplifiers' input. Higher values of R_L will not appreciably increase the measured signal. In the other direction, assuming an amplifier with very high input impedance, for a given light level, R_L must not be so high that saturation occurs. Hence, it is often convenient to employ a variety of values for R_L, as shown in Fig. 8. Thus, at high light levels, one can use lower values of R_L to prevent saturation, and at low light levels, increase R_L to increase the measured signal level. This will increase the dynamic range over which signals can be measured. The amplifier used should have an input impedance at least 100-fold greater than the highest value of R_L, to prevent measurement error. Frequently, a selector like that shown in Fig. 8 is part of the amplifier input circuit and need not be constructed. Often the amplifier input will employ a voltage divider (similar to that shown in Fig. 9) rather than different R_L's so that the input impedance remains constant. Essentially the same effect results, although there may be a disadvantage in this latter technique in terms of noise. The source noise will increase as $(R_L)^{1/2}$, while the signal of interest increases as R_L (at constant light flux or average anode current). Therefore, one should use a value of R_L as large as possible (without causing saturation) which will improve the signal-to-noise ratio (S/N) across R_L. Also, the photomultiplier can be operated at lower gain, due to the larger signal available.

This is advantageous, since the S/N for photomultipliers usually deteriorates at high gain (high overall photomultiplier voltage).

Quite often one hears of amplifiers or similar measuring devices referred to as *electrometers*. An electrometer is simply an amplifier of unusually high input impedance. It is not quite clear just when it ceases to be an "amplifier" and becomes an "electrometer." Some examples: any device capable of measuring extremely small currents, say below 10^{-10} A, or capable of measuring voltages across very high resistances, say 10^8 Ω or more, would probably be called an electrometer.

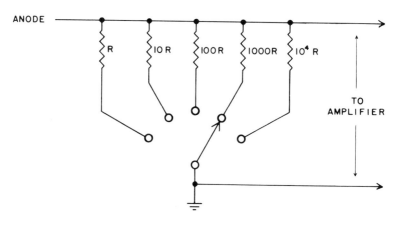

Fig. 8. Multiple load resistor selector for decade changes in R_L.

C. AMPLIFIER OUTPUT

The amplifier output is usually a dc voltage appearing across some relatively low amplifier output impedance. This is usually connected across the input of some readout device, such as a recorder, having a considerably higher input impedance. Normally, this presents little trouble, provided the proper input-output resistance relationship is maintained. Let us look at one example of a possible situation that might occur. You have an amplifier, the input of which is connected to your photomultiplier. The output terminals of the amplifier produce 10 V for a full-scale meter reading, and the output impedance of the amplifier terminals is 10 kΩ. Suppose one wishes to connect this to a strip-chart recorder, but the only one available has a full-scale sensitivity of 10 mV, and a maximum measuring resistance tolerance of 5 kΩ. To connect these, you will have to employ a voltage divider network, such as that shown in Fig. 9. Here one can readily see that the current flowing through the 100-kΩ and 100-Ω resistors is

$$i = E/R = 10/100 \text{ k}\Omega \simeq 0.0001 \text{ A} = 0.1 \text{ mA}$$

This will produce a voltage drop across the 100-Ω resistor of $(0.1)(100) =$ 10 mV, or the required 1/1000 of the amplifier output. Also, the resistance across the recorder input is much less than the maximum allowable 5 kΩ. Of course, other resistance values could be used, but their ratio must be

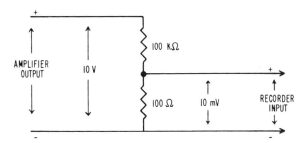

Fig. 9. Voltage divider.

1/1000. They must also not be so large that the resistance across the re-corder input exceeds 5 kΩ, nor so small that excessive current is drawn from the amplifier output terminals. Also observe that the proper polarity must be maintained (positive-to-positive).

D. FREQUENCY RESPONSE AND NOISE

Before considering frequency response and noise in amplifiers, let us place several restrictions or assumptions on the system. First, the system to be discussed employs a photomultiplier detector. Second, the source radiation is modulated, resulting in an ac signal. Third, an ac amplifier is used, which does not respond to dc (i.e., a blocking capacitor is used in the input). The first is justified, since most spectrometers employ photo-detectors. The second and third rule out dc measurements with chopper-stabilized dc amplifiers, or electrometers. This is probably not justified, since many instruments do employ such a system. However, they have one disadvantage compared to ac amplifiers, the so-called "flicker noise," or $1/f$ noise. At frequencies (f) less than about 100 Hz, the $1/f$ noise usually dominates thermal or shot noise and is troublesome when trying to measure very small dc signals. Flicker noise is not completely understood but appears to arise from statistical fluctuations in electrical conduction through or electron emission from electronic devices, such as vacuum tubes and transistors. The noise power (per cycle) appears to follow a $1/f$ distribution below about 100 Hz and implies that dc (i.e., 0 Hz) is the worst possible measurement frequency to use, at least from a S/N standpoint. Therefore, if one modulates the source radiation, generating his signal at a higher frequency, and uses an ac amplifier, it is possible to minimize the $1/f$ noise. One should always avoid dc measurements when S/N is a problem.

Assuming now that we are chopping our source at some frequency above 100 Hz and using an ac amplifier, we have more or less eliminated two sources of noise: drift and $1/f$ noise. This leaves us with shot noise (at moderate light flux) and Johnson noise (at very low light flux.) Let us assume that both of these are "white" noise, i.e., noise power vs frequency is uniform. Note that dark-current noise is not important here, since we are using modulation and ac amplification. Noise in the amplifier itself can also be important, but usually at high light flux levels only (7). One very important consideration is the frequency response bandwidth Δf of the system.

In atomic absorption spectrometry, for example, the sensitivity is determined by the slope of the absorbance vs concentration curve. Since the absorbance A is given by log I_0/I, increasing the source intensity (I_0) will *not* increase the sensitivity because the ratio I_0/I will remain unchanged. However, the detection limit (i.e., minimum detectable concentration) can be improved because the greater source intensity will allow reduced amplifier and/or photomultiplier gain, thereby reducing the noise level. Thus, one obtains with a more intense source an improved S/N ratio: same signal, less noise. This improves the detection limit, which is defined in terms of S/N.

Another means of improving the S/N ratio is to decrease the frequency response bandwidth (Δf) of the system. Our signal, let us say at 330 Hz, is superimposed on some level of white noise. If we use a broad-band ac amplifier, e.g., one with a response from 10 Hz to 1000 Hz, the signal will be amplified. But, all of the noise within 10 and 1000 Hz will also be amplified. As we go to smaller and smaller signals, we reach a point where the signal becomes indistinguishable or buried in the noise. If we now modify our amplifier so that it responds only between 100 and 500 Hz, the signal will still be amplified, but less of the noise will be amplified. If the noise is "white," then less of it is present in the Δf of 400 Hz (between (100 and

500 Hz) than in the previous Δf of 990 Hz (10 to 1000 Hz). One can continue this process of narrowing Δf and improving the S/N ratio. There are two widely used techniques for accomplishing this: tuned ac amplifiers and synchronous amplifiers (or "lock-in" amplifiers).

E. TUNED AC AMPLIFIERS

Most ac amplifiers usually employ some type of negative feedback to stabilize gain, reduce noise, and increase the input impedance. Using a *tuned* network in the feedback loop, it is possible to alter the frequency response of the amplifier, making it considerably more frequency selective.

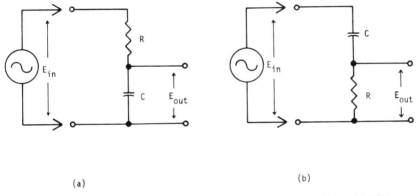

(a)   (b)

Fig. 10. *RC* filter networks (ac voltage dividers): (a) low-pass filter; (b) high-pass filter.

First, let us look briefly at ac voltage dividers (filters) composed of resistor-capacitor (*RC*) combinations. Two types of *RC* filters are shown in Fig. 10. The circuit shown in Fig. 10(a) is called a *low-pass filter* because E_{out}, the voltage across the capacitor, decreases as the frequency increases (assuming a constant E_{in}, of course). Thus, impedance of the low-pass filter increases as the frequency increases, so only low frequencies are "passed." The upper cutoff frequency is taken as that frequency where $E_{out} = 0.707E_{in}$. The value 0.707 is $\sqrt{2}/2$, or the root-mean-square (rms) value rather than the peak value of the voltages. It can be shown (6) that the upper cutoff frequency f_u is given by

$$f_u = \frac{1}{2\pi RC} \tag{2}$$

This is simply the frequency at which the capacitor's impedance is equal to R.

For the high-pass filter, Fig. 10(b), the voltage across the resistor will decrease as the frequency decreases. The reason is that, as the frequency decreases, the capacitor's impedance increases and the fraction of E_{in} found across R decreases. Therefore, only high-frequency signals will be "passed" unattenuated. Again, the frequency at which the capacitor's

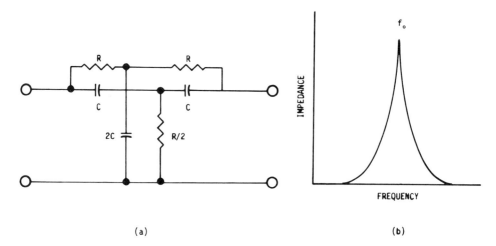

(a)   (b)

Fig. 11. Twin-T filter: (a) schematic with characteristic frequency, $f_0 = \frac{1}{2}\pi RC$, (b) graph of impedance vs frequency.

impedance is equal to R is called the lower cutoff frequency f_L and is identical to that of Eq. (2):

$$f_L = \frac{1}{2\pi RC} \tag{3}$$

Hence any dc component in the signal will not be passed by C and thus will not appear across R.

These two filters can be combined into one filter with relatively narrow frequency response. One common arrangement, shown in Fig. 11(a), is called a twin-T filter. It is composed of a low-pass filter (R, R and $2C$) and a high-pass filter (C, C and $R/2$); the impedance [Fig. 11(b)] has a sharp maximum of f_0. If this filter is incorporated into the feedback loop of an ac amplifier (Fig. 12), the result is an amplifier having maximum response at f_0, i.e., a tuned amplifier. The reason is as follows. Negative

feedback decreases the amplifier gain. The twin-T has maximum impedance at f_0; therefore, the feedback is a minimum (thus, the gain is maximum) at f_0. Hence, one obtains a rather sharply tuned ac amplifier
which responds only to a narrow band of frequencies around f_0 (i.e., Δf
small). Also, the degree of rejection of frequencies other than f_0 increases
as the amplifier gain increases.

If one uses several successive stages of tuned amplification, all tuned
to the same frequency, the frequency band width (Δf) of the overall system

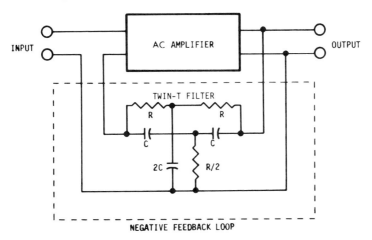

Fig. 12. Tuned ac amplifier.

is made narrower and narrower, improving the S/N ratio. It is possible to
make tuned ac amplifiers (having several stages) with a Δf of only a few
Hz. However, this presents an alignment problem, because changing
(peaking) the tuning on one stage upsets the tuning on the other stages.
Also, due to the sharp maximum in the response [Fig. 11(b)], any slight
drift in the signal frequency (i.e., the chopping frequency, in our example)
will cause serious changes in the signal level shown at the readout. There
is a better way to amplify ac signals and achieve a very narrow Δf, without
drift problems, and achieve excellent S/N ratio (up to the theoretical maximum). This is accomplished with a phase-sensitive or "lock-in" amplifier.

F. LOCK-IN AMPLIFIERS

There is nothing really complicated about the principle of operation of
a lock-in amplifier. A block diagram of a typical one is shown in Fig. 13.

Before discussing the operation of a lock-in amplifier, let us discuss a typical application. Figure 14 shows a typical setup for atomic fluorescence spectrometry employing a lock-in amplifier. The chopper modulates the radiant energy source. Therefore, the fluorescence, and hence the signal, will appear at the chopping frequency. The reference is modulated by the same chopper and will therefore have exactly the same *frequency* as the desired signal. Any signals at any other frequency are undesirable, since they will only contribute to the overall noise.

Looking again at Fig. 13, the signal is amp tied by an amplifier to raise its power level. This ac amplifier may or may not be tuned to the chopping

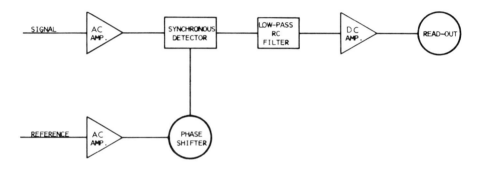

Fig. 13. Block diagram of synchronous or lock-in amplifier.

frequency, but usually is moderately tuned to eliminate much of the noise (i.e., signals at other frequencies, which are simply noise) at frequencies far removed from the chopping frequency. The signal is then fed into the synchronous detector, along with the reference. The synchronous detector can be thought of initially (but erroneously) as a "mixer," which simply beats the signal and reference together. The resultant (output) consists of *sum* and *difference* frequencies, the sum being a high frequency (e.g., twice the signal frequency) and the difference being a low frequency (i.e., 0 Hz, or dc). The low-pass filter allows only the difference frequency (dc) to pass and blocks or throws away the higher sum frequencies. Now the dc signal is amplified by a dc amplifier and fed to some readout device, such as a potentiometric recorder. The *effective noise bandwidth* (i.e., Δf) of the amplifier is determined by the frequency response (above 0 Hz) of the low-pass RC filter. The frequency response of the low-pass filter is usually expressed in terms of its *risetime* τ, in seconds, where τ is the time required for the output level to go from 10% to 90% of its final value. The relationship is as follows:

$$\text{equivalent noise bandwidth} \simeq \frac{1}{10 \text{ to } 90\% \text{ risetime}} \qquad (4)$$

For amplifiers employing double section RC filters, the 10 to 90% risetime is approximately equal to $4RC$. Thus, the effective Δf of the amplifier is

$$\Delta f \simeq \frac{1}{4RC} \qquad (5)$$

For example, if the low-pass filter has a time constant (RC) of 1 sec, the effective Δf of the amplifier is 1 Hz. If the time constant is increased to

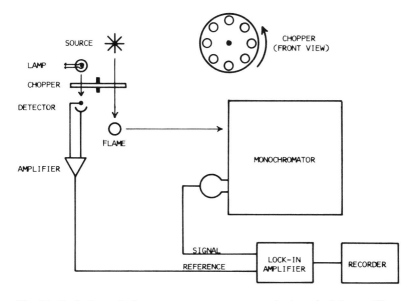

Fig. 14. Typical atomic fluorescence spectrometer employing a lock-in amplifier.

10 sec, the Δf is 0.1 Hz. Thus, one can arbitrarily decrease Δf of the amplifier, but in doing so, one must sacrifice time. In the above example, changing Δf from 1 Hz to 0.1 Hz, one has to wait 10 times longer for a response.

It should now be apparent that lock-in amplifiers are an excellent way to obtain very narrow effective noise bandwidth and, consequently, of improving the S/N ratio of the measurement. One can trade time for S/N

improvement. Unfortunately, the trade is not linear. The noise decreases only as the (time)$^{1/2}$. For example, going from a time constant of 1 sec to one of 100 sec narrows Δf by a factor of 100, but this only decreases the noise by $(100)^{1/2}$ or 10. Thus, you have to wait 100 times longer to improve your answer by a factor of 10. This is the theoretical maximum rate of improvement. No other system or technique will improve the S/N ratio faster (in terms of time) than as a square-root relationship. Therefore, this is a theoretically optimum system, in terms of S/N ratio.

The synchronous detector mentioned earlier in Fig. 13 could be thought of (erroneously) as a mixer circuit which beats the signal and reference together. A mixer circuit would require that the magnitude (e.g., voltage) of the reference vary in the same manner as the signal and not change appreciably with time. In practice, lock-in amplifiers do not employ such a mixer as the synchronous detector, but rather use an electronic switch (often a type called a "Schmitt trigger"). The Schmitt trigger is simply an electronic switch (double-pole, double-throw) which is driven by the reference signal at the reference frequency. Only those signals with the same frequency (and phase) as the reference will pass through the switch unattenuated. This greatly lessens the requirements on the reference; it need only be of sufficient power to operate the switch, not necessarily stable. In such an arrangement, the signal and reference must have the same frequency, and in our example shown in Fig. 14 they will have exactly the same frequency, since both are modulated by the same chopper. The signal and reference must also be in phase, and hence the phase shifter shown in Fig. 13. This allows one to shift the relative phases of the reference and signal so as to obtain maximum response. In our example (Fig. 14), the phase shifter is not an absolute necessity since the phase could be adjusted by moving the reference generating system to a suitable position around the chopper. Another advantage of the lock-in amplifier is that small changes in the chopping frequency will not greatly change the signal level, since both the reference and signal frequency will still be the same. Also, the lock-in amplifier is finding widespread and ever increasing use in flame spectrometry. It is quite likely that the better instruments (emission, absorption, and fluorescence) in the future will employ exclusively this type of amplifier, largely replacing the tuned ac, ac and dc amplifiers in use today.

Another consideration might be the selection of the chopping frequency. In general, frequencies below 10 Hz are to be avoided, because of the $1/f$ noise mentioned earlier. Also, the frequency of the power line (60 Hz in the U.S.A.) and its harmonics are to be avoided (e.g., 120, 180, 240, etc.).

There harmonics decrease rapidly above the fourth or fifth, making the region from about 300 Hz all the way up to the AM radio band ($>10^5$ Hz) a relatively "quiet" region. Using mechanical choppers, one is generally limited to a modulation frequency of less than 1000 Hz. Mostly for convenience, choppers with 11 or 13 apertures driven by an 1800 rpm synchronous motor have been used by the author. These provide chopping frequencies of 330 or 390 Hz, respectively, which are between the fifth and sixth, and sixth and seventh line frequency harmonics, respectively. The fifth, sixth, and seventh harmonics are apparently very weak, and, as these chopping frequencies are 30 Hz away from the harmonics, no pickup has been noted when using lock-in amplifiers.

G. Signal Averagers

Another type of electronic device for obtaining optimum S/N ratio improvement, which has not found widespread use in flame spectrometry, is the signal averager. There are several types of these, both analog and digital, with the latter the more popular at present. They are given various names, such as "computer of average transients," "digital memory oscilloscopes," "boxcar integrators." Essentially, they take many "slices" of a repetitive signal, usually buried in noise, and "add" or average each slice into their respective channels or capacitors. The noise, if truly random, should add to zero, while any signal present would not. The slices, after some period of time (let us say in digital form), are converted to analog and displayed on some suitable readout device, such as an oscilloscope. The conversion to analog is not always necessary, depending on the particular instrument. Compared to lock-in amplifiers, they are considerably more expensive. Also, there is no time advantage in improving S/N. The noise adds as the square root and the signal adds arithmetically. After 100 repetitions of the waveform, the signal is increased 100-fold and the noise by $(100)^{1/2}$. Therefore, the S/N ratio is improved by a factor of only 10, just as with the lock-in amplifier.

V. Readout Devices

A. Meters

By far the most popular readout devices used at present in flame spectrometry are meters and recorders. Galvanometers are seldom used anymore, except in conjunction with some of the simpler filter instruments which do not employ amplifiers. The galvanometer readout is connected directly to the detector, such as a barrier-layer cell. They are also used in densitometers, in conjunction with photographic detectors, although this detector is rarely used in flame spectrometry. Technically, the galvanometer is simply a meter with very good current sensitivity. Almost all meters used as readout devices are of the moving-coil or *D'Arsonval* type, and measure dc current. They have a constant resistance, so that the current through the meter is proportional to the voltage across it. Hence, they are usually measuring the dc output of the amplifier. It is of little purpose here to go into the intricacies of meter design and operation. Those interested further in this subject will find a good discussion in Ref. (6).

B. RECORDERS

Recorders vary widely in design, from galvanometers with a pen attached to their indicating arm to ones with linear servo motors and even without servo motors. However, most are of the potentiometric type, which utilize a servo motor to balance a potentiometer bridge. Most recorders employ an ac two-phase induction servo motor to move the recording device, usually an ink pen, across the width of the chart. Displacement is proportional to the dc signal input, although some recorders employ a logarithmic slide wire or gears to achieve a displacement proportional to the logarithm of the input. This is often useful in recording absorbance directly, as in atomic absorption. However, most atomic absorption instruments will employ, instead, a logarithmic amplifier so that ordinary linear recorders can be used. Output terminals for a recorder are available on most metered instruments. Many recorders in use today have multiple input ranges (1 to 10 mV) achieved by switching in various voltage dividers on the input. Single range recorders can be used, provided that the amplifiers' output (full-scale) is equal to or greater than the recorders' span. If greater than, the meter and recorder can be made to correspond by using a suitable voltage divider network between the amplifier output and recorder input. Most recorders used are of the "strip-chart" or "*T-Y*" variety. In this type, the pen displacement (y axis) is determined by the signal magnitude while the chart is driven at a constant speed. Hence the x axis is simply time. Some recorders are designed so that the x axis is displaced by another signal; hence, one obtains a plot of one signal against the other. These are usually referred to as X-Y recorders, but used infrequently with flame spectrometers, the strip-chart variety being by far the most popular. Those interested in the more technical aspects of recorder design are referred to Ref. (6) for additional information.

Recorders have several advantages over meter readouts. Meters can be misread and the number written down will be in error. A recorder provides a permanent record of the measurement which allows one to recheck the data.

Probably the biggest advantage of recorders over other readout devices is that one has a *continuous* record of the measurement and can accurately determine the magnitude of any noise present. Thus, one can note any sudden or subtle changes in the signal, and accurately determine the S/N for his measurements. This is important since it is usually the S/N that determines the detection limit in flame spectrometry, and one usually wishes to optimize the S/N ratio and not the signal alone. When scanning the wavelength with a spectrometer, the recorder is certainly the most suitable type of readout device.

C. DIGITAL AND ANALOG DEVICES

Another type of readout device gaining in popularity, especially in commercial atomic absorption instruments, is the digital readout. Some are truly digital, others are pseudodigital readout devices. The latter type are really analog devices, just as meters and recorders are, and the analog mechanism (e.g., an analog servo mechanism) drives a pseudo digital counter, much like the odometer on an automobile. Most of the presentation is "digital," except perhaps for the last significant figure or two. The truly digital readout is one that converts the analog signal to a digital signal (e.g., a number of pulses) and displays the digital signal on a non-analog counter or lighted number display, such as "nixie" tubes. Hence, there is never any ambiguity as to the numerical value of the signal, since only one answer is presented. Interpolation between numbers or marks never has to be made.

In digital instruments no "operator bias" will enter the readings. No interpolation or rounding-off decisions need to be made; only one value is presented. The advantage of truly digital readout devices is that they employ a decimal-to-binary converter (binary coded decimal or *BCD* output) on the output; the signal is fed into a printer and/or tape punch for subsequent computations on a digital computer. With the numerical value unambiguously printed out, one combines the advantages of digital readout with one advantage of recorders, namely, a permanent record. With the punched tape, one can also read his data directly into the computer and obtain virtually any degree of interpretation or presentation desired; concentrations can be calculated, working curves plotted, statistical analyses performed, data tabulated, final report typed, etc. This also has the advantage that the data is not "handled" by the operator, with no risk of copying down or punching in a wrong number. The results are "untouched by human hands," so to speak.

The greatest application of these read-out devices in flame spectrometry has been in atomic absorption spectrometry, particularly the more sophisticated commercial instruments designed to handle large numbers of samples, operated by relatively unskilled personnel, or highly automated systems. One popular unit for atomic absorption spectrometers converts directly the amplifier's recorder output to a digital readout, in terms of percent transmittance, absorbance, or even concentration. It also employs S/N ratio improvement by means of signal averaging and has provisions for correcting working curves which deviate from Beer's law (nonlinear). Provisions for scale expansion are incorporated, and there is also a BCD output, for connection to printers, sequencers, paper tape punches or magnetic tape recorders, for subsequent data processing.

REFERENCES

1. K. S. Lion, *Instrumentation in Scientific Research*, McGraw-Hill, New York, 1959.
2. R. D. Sard, *J. Appl. Phys.*, **17**, 768 (1947).
3. R. W. Engstrom, *J. Opt. Soc. Am.*, **37**, 420 (1947).
4. S. Rodda, *Photoelectric Multipliers*, McDonald, London, 1953.
5. R. W. Engstrom and E. Fisher, *Rev. Sci. Instr.*, **28**, 525 (1957).
6. H. V. Malmstadt, C. G. Enke, and E. C. Toren, Jr., *Electronics for Scientists*, Benjamin, New York, 1963.
7. J. D. Winefordner and C. Veillon, *Anal. Chem.*, **37**, 416 (1965).

7   Instrument Operation

C. Veillon

DEPARTMENT OF CHEMISTRY
UNIVERSITY OF HOUSTON
HOUSTON, TEXAS

I. Introduction

We will consider here general procedures and precautions for the various commercial instruments. Although each instrument has differences in operational procedure, they are all basically alike in principle. Some of the parameters to be covered are not applicable to certain instruments because no provision was made by the manufacturer to allow the operator to adjust, monitor, or optimize a particular parameter. This may be simply a case of poor design or deliberate cost reduction on less sophisticated instruments. For example, many instruments have no provision for monitoring the flow rate of gases going to the burner, only their pressure. As will be shown later, it is quite possible for the flow rate of a gas to change without the pressure changing. As another example, some instruments employ fixed slit widths on the monochromator which limit the versatility of the instrument and increases the chance of error.

II. Preliminary Operations

With a new instrument, STUDY the instruction manual before attempting any operations or adjustments. The basic operations to be described can be applied generally to any instrument; however, some modifications may be necessary and the instrument instruction manual should be consulted for the exact details. The following operations are discussed but not in any particular or preferred order.

A. Source Stabilization

For atomic absorption and atomic fluorescence measurements, an external source is used and must be allowed to operate for a time period sufficient to stabilize its output. Intrinsic source stability is discussed in Chapter 2 of this volume. Light sources, such as hollow-cathode lamps, vapor discharge lamps, and electrodeless discharge tubes all have a finite "warm-up" time before their light output becomes essentially constant. In atomic absorption with single-beam instruments, this lamp stabilization is most important since drift in the source intensity during a measurement will result in an error. It is less of a problem in absorption measurements with double-beam instruments since the ratio of the sample and reference beams should remain constant even if the source intensity drifts. Most sources take about 15–30 min to stabilize completely, so the double-beam instrument has a time advantage if one needs to rapidly change from one source to another. Usually this problem is overcome in single-beam instruments by mounting several sources on a rack or turret in such a manner that all are "warmed up" at the same time. To change sources, one then simply moves the next source (already stabilized and operating) into position. When possible, it is usually best to let the source stabilize completely before using, even in the double-beam system. As will be shown later, it is possible for the optimum source intensity and monochromator spectral slit width to change while the source is stabilizing.

Hollow-cathode discharge lamps are widely used as primary sources in atomic absorption flame spectrometry. Most of the commercially available lamps today are designed so as to be quite stable in spectral output, both short-term and long-term. This assumes a power supply of sufficient stability. Normally one would use a current-regulated power supply to power the hollow-cathode lamp. This power supply would employ feedback so as to stabilize the current output of the supply. This is usually more satisfactory in terms of stability than using a voltage-regulated power supply, since it is the current through the lamp that has the greatest effect on the output intensity.

In atomic fluorescence spectrometry, one usually employs electrodeless discharge tubes (microwave discharge) as the primary source, or continuum sources. Usually, source instability is less of a problem in atomic fluorescence than in atomic absorption, especially at low concentrations and near the limit of detection. As one approaches the detection limit, the signal becomes a *maximum* in atomic absorption, but becomes a *minimum* in atomic fluorescence. Hence, near the detection limit, any source instability will have a maximum effect in absorption and a minimum effect in fluorescence. At high concentrations, the opposite is true.

B. FLAME STABILIZATION

This is one of the more important aspects of flame spectrometry, but many instruments do not have the means to do this properly. The most detrimental type of flame instability is caused by changes in the gas flow rates, especially the aspirating (nebulizing) gas, and the fuel under fuel-rich conditions. The former affects the amount of sample reaching the flame and the latter affects the number of atoms produced in the flame. Any temperature fluctuations will have a serious effect on flame emission and a lesser effect on atomic absorption or fluorescence. Emission by atoms increases exponentially with temperature, whereas absorption or fluorescence vary only as the square root of the temperature. The most critical parameter in flame stability is control of the gas flow rates.

Another type of "flame instability" that may be encountered is that due to thermal heating of the instrument and/or components by the flame itself. It is possible for temperature changes in the burner to alter its dimensions and, therefore, its characteristics. However, this is usually a minor problem, if present at all. A more frequently noticed problem is that of "wavelength drift." Often one will peak the wavelength setting of the monochromator on the desired wavelength. Radiation and/or conduction of heat from the flame will gradually increase the temperature of the mechanism in the monochromator, causing the wavelength setting to drift off its peak setting. The two most common ways of overcoming this type of instability are to allow the instrument to "warm up" with the flame on, say for 30 min, until it comes to thermal equilibrium before peaking the wavelength setting, or to take readings by scanning over the desired wavelength. The latter is applicable only to instruments that employ scanning monochromators and recorder readouts. This would be the procedure to use in any case when maximum precision is required. Another suitable method would be to use an exit slit width slightly larger than the entrance slit width on the monochromator. However, few instruments have separately adjustable entrance and exit slits, so this technique is not usually employed. Bear in mind that this thermal drift of wavelength setting is not really an instability of the flame. It is simply a result of the heat from the flame and is only a problem when the instrument's temperature is changing, usually in the first 30 min or so of operation.

In this connection, it should also be pointed out that a reasonably constant temperature is required in the laboratory for the problem mentioned above, and more important, to prevent temperature (and therefore, viscosity) changes in the sample solutions and standards.

As mentioned earlier, it is the gas flow rates which are the most important parameters in operating a flame in a constant and reproducible manner. It is relatively common practice (or "malpractice") to monitor only the *pressures* of the gases fed to the burner. This is fine, and works satisfactorily until one encounters partial clogging in the burner or nebulizer. It is quite possible for the gas flow rate to change in many instruments without a pressure change. Consider one typical, poorly designed setup. The gas pressure from the cylinder is regulated by a good two-stage regulator. On the instrument there is a *pressure* gauge to monitor the pressure of the gas going to, let us say, the aspirator. If this gas line now becomes partially clogged, the *flow* rate of gas to the aspirator will be reduced, but our good two-stage regulator will continue to maintain *the same pressure* in the gas line, and the gauge on the instrument tells us nothing. A much better arrangement would be to monitor the flow rate of the gases, rather than their pressure. This is most important, as trouble is indicated immediately. The better instruments that do employ gas flow measurement (or indication) usually employ one of two methods: the differential pressure method or the rotameter method. The differential pressure method is illustrated schematically in Fig. 1. This illustrates a manometer-type pressure gauge, although a Bourdon-tube type pressure gauge could also be used. The Venturi causes a pressure drop, which depends on the gas flow rate.

A more popular flow measuring device is the rotameter tube flowmeter, as illustrated in Fig. 2. There are many variations with various float shapes and materials, but they all work on the same principle. The one illustrated in Fig. 2 employs two floats of different density to increase the dynamic range of flow measurement. Although they can be obtained calibrated

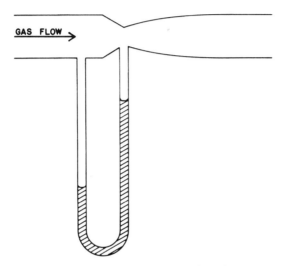

Fig. 1. Schematic of differential pressure method of gas flow rate measurement.

directly in flow rate of a specific gas, most are graduated in arbitrary units. Calibration curves for various gases at specified temperatures and pressures are usually available from the manufacturer, and are good to within a few percent. They can be calibrated by the user, using an inverted, water-filled graduated cylinder, or better, using a wet test meter with the metering valve downstream from the rotameter. If pressure-regulated gas is fed to the rotameter, this insures that the gas pressure (and, therefore, its density, viscosity, and so forth) is always the same in the rotameter, so that the calibration is constant. Otherwise, it will be necessary to make corrections (*1*). When calibrating the rotameters, the same part of the float (e.g., bottom, middle, top) should be used for all readings, and for all subsequent readings. In terms of flame stability and reproducibility, the author recommends the type system shown in Fig. 3 as the *minimum* requirement. More gases can be added if needed in the particular system.

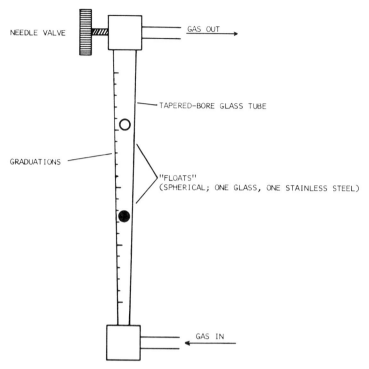

Fig. 2. Schematic of a rotameter gas flow measuring device.

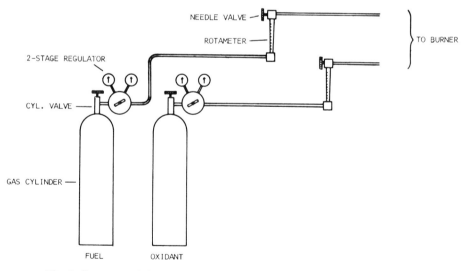

Fig. 3. Recommended minimum gas handling system for flame spectrometry.

182

C. DETECTOR STABILIZATION

Normally, this presents no serious difficulties. Most instruments today employ photomultiplier detectors, which are quite stable. A good, high-stability high-voltage power supply should be used with these detectors, since any drift in the applied voltage will affect the overall response. The minimum recommended electrical specifications for this power supply were outlined in Chapter 6. The gain stability of photomultiplier detectors is quite good, so long as one is careful in preventing fatigue (see Chapter 6). After exposure to excessive light, they may become somewhat "noisy," but will usually recover after a period of darkness.

III. Optimization

Before performing an analysis, be it by flame emission, absorption, or fluorescence techniques, there are several important decisions to be made. These include the selection of the line or wavelength, the slit width, the fuel, oxidant and their flow ratio, and the type of burner and sample introduction system to be used. These parameters are not all independent. Also, one may not be able to optimize all of these for obvious reasons, such as only one burner is available, and fixed (nonadjustable) slit width. Assuming that we have a versatile instrument and the necessary accessory equipment and supplies, let us look briefly at how one goes about selecting and optimizing these parameters.

A. WAVELENGTH SELECTION

The selection of a wavelength (line) for a flame analysis is often done rather haphazardly. Usually one is seeking the most sensitive line for an analysis and assumes that the one often used by others, or the one listed in the instruments' instruction manual is the most sensitive. Usually, it is. Sometimes one is trying to find a line that is less sensitive than the best one, to extend the dynamic range of the analysis to higher concentrations without having to dilute the sample. One would then like to have some idea of, or be able to predict roughly, the relative sensitivities of other lines of the element in question. Some instrument manufacturers provide these data with their instruments. Alternatively, one can measure the relative sensitivities for himself. However, it is *usually* possible to predict the most sensitive line, and the very approximate relative sensitivities of the other possible lines, *for atomic absorption and atomic fluorescence.* This is not easily accomplished for atomic emission, because of the limited energy available in the flame for excitation. Going to a flame of different temperature may alter the choice of the most sensitive line in many cases.

First, the selection system used by the author will be described and will be followed by some illustrations of its drawbacks and limitations. The source of information is the tables of transition probabilities by Corliss and Bozman (2). These tables give the wavelength of the lines, the energy levels of the transition (in kaysers, or cm^{-1})* the gA and gf values. The gA value is the product of the statistical weight (upper level) and the Einstein transition probability. The gf value is the product of the statistical weight and the oscillator strength. In selecting the most sensitive line for atomic absorption or fluorescence, it is *usually* that wavelength with the greatest gA or gf value *and* originating in the ground state (0 cm^{-1}). For example, the copper transition at 3247.54 Å originates in the ground state (0 → 30,784 cm^{-1}) and has the largest gA value (and gf value) of all atomic copper transitions *originating in the ground state.* Thus, one would predict that this line would be the most sensitive one for atomic absorption or atomic fluorescence, and it is. The transition with the largest gA value (and gf) is the 5218.20-Å line (30,784 → 49,942 cm^{-1}), but note that the initial state (30,784 cm^{-1}) is about 3.6 eV above the ground state. Thus, one would not expect any measurable degree of sensitivity because virtually none of the free atoms would be in this level initially.

Using this rather simple method, one can usually select the most sensitive line of an element for atomic absorption or fluorescence by consulting the data in the Corliss and Bozman tables (2). This is often helpful with elements having large numbers of lines, like Fe, Ni, and Co. There are two things that one must keep in mind when using this simple technique: (a) there are exceptions, and (b) the technique as outlined is not reliable in accurately predicting the *relative* sensitivities of the various lines.

This technique cannot be so simply applied to atomic emission, unless the intensity is calculated (2) from particle density, partition function, and temperature data. A quick procedure one might use to determine the most sensitive line of an element for flame emission is to measure the relative sensitivities of promising lines above about 2500 Å. This wavelength corresponds to an energy of 5 eV, about the analytically useful

---

* 1 eV is equivalent to 8065.9 kaysers.

limit for most chemical flames. An even quicker method is to look up the relative emission sensitivities tabulated by Gilbert (3) for various flames. or consult Volume 3 of this series.

B. SLIT WIDTH

The monochromator entrance and exit slit widths, usually the same in most instruments, determines the spectral bandpass of the instrument, as well as the light flux reaching the detector. Selecting the optimum slit width for a given analysis is often of critical importance in flame emission, less so in atomic absorption and even less so in atomic fluorescence. The optics involved in determining the spectral bandpass for a given instrument have been described in Chapter 5, so we will concern ourselves here only with optimization of the spectral bandpass by slit width adjustment, assuming that both the entrance and exit slits are identical and adjusted bilaterally. In many of the simpler instruments, fixed slits are employed. This eliminates the problem of having to optimize the spectral bandpass, on the other hand you can rarely operate the instrument under optimum conditions.

Several workers have investigated the selection of optimum conditions for an analysis, from both an experimental and a theoretical point of view. Cellier and Stace (4) have developed a statistical approach for optimizing the various operating parameters, although they did not consider slit width *per se*. Strasheim and Wessels (5) examined the effect of slit width and other parameters in the determination of some noble metals. Winefordner (6) has considered the effect of slit width on intensity of flame atomic emission and on atomic absorption. He found that, ideally, the observed intensity in emission should increase directly as the square of the slit width. Thus, when doubling the slit width, one can expect the measured intensity (line) to increase by a factor of four. Note that this is for the ideal case. Flame background and nearby lines will also have to be taken into account in a practical analysis. In atomic absorption spectrometry, one usually employs a line source in which the source emission line width is narrower than the absorption line width of the atoms in the flame. This is necessary to assure linear analytical calibration curves and maximum sensitivity. In this case, the effective spectral slit width of the monochromator is determined by the source emission line width. The monochromator is used simply to isolate the desired line from the other source emission lines. Here one would adjust the monochromator spectral slit width to as wide a value as possible without taking in other source lines. This will maximize the source intensity (I_0), which increases as the square of the slit width. Maximum source intensity is often desirable in improving the signal-to-noise ratio, as explained in Chapter 6.

In most flame methods, be it emission, absorption, or fluorescence, it is the signal-to-noise ratio that must be optimized, and not just the signal alone. Since the signal-to-noise ratio and the signal are not usually independent, the optimization process is not quite so simple. Methods of optimization of the various important parameters, including slit width, are discussed by Parsons and Winefordner (7). Winefordner and Vickers (8) have derived expressions for calculating the optimum slit width (for emission) in terms of other instrumental parameters. Examples are given, and it is interesting to note the considerable differences in optimum slit widths with different instruments. No doubt this would be a worthwhile calculation to make on the specific instrument that one is using in an analysis.

The references mentioned in the above paragraphs should provide one with a means of calculating and/or measuring the optimum slit width for an analysis. As considerable material is covered in these references, it will not be reproduced here.

In general, the slit width should be adjusted to the point that gives the greatest signal-to-noise ratio. For flame emission (atomic line) this may be a fairly wide slit where the flame background emission is low, or a very narrow slit where the background emission is high. In atomic fluorescence, one is usually measuring the fluorescence in a flame region of very low background emission, so very wide slits are employed. Here, too, the signal-to-noise ratio must be maximized, and not just the signal. In atomic absorption the source emission line width should be (and usually is) narrower than the atoms' absorption line width in the flame. As pointed out earlier, this results in linear working curves and maximum sensitivity. Here the effective spectral bandwidth is that of the source emission line width, and it is often necessary to use wide slit widths to maximize the signal-to-noise ratio, but not so wide as to take in other source emission lines. Often, when the source has a complex spectrum and/or lines very close to the desired line, very narrow spectral slit widths must be employed. Otherwise, the nearby (unabsorbed) source lines will be detected, reducing the sensitivity. If a scanning monochromator is employed in atomic absorption, or if a spectrum of the source is available, one can determine the spectral slit width to be used. It is simply the widest that can be used without taking in other source lines or radiation.

For atomic absorption, there are two cases which have not yet been mentioned. The first is the case where the source emission line width is greater than the atoms' absorption line width in the flame. This might occur when the source (e.g., a hollow-cathode lamp) is run at an excessive current, so that the source emission line is broadened. Again, the effective spectral bandwidth will be essentially determined by the source emission line width, if we assume that the spectral bandwidth of the monochromator is greater than the source emission line width. In this case, unabsorbed radiation will be detected, lowering the sensitivity. The second case not yet mentioned is when a continuum source is employed. Here the effective spectral bandwidth is that of the monochromator itself and if this is greater than the absorption line width, unabsorbed radiation will be detected, lowering the sensitivity. This is almost always the case with the low-to-medium resolving power monochromator employed in atomic absorption and, for this reason, continuum sources are rarely used in atomic absorption spectrometry. Monochromators with sufficiently narrow spectral bandwidth could overcome this limitation, but their cost would negate the advantage that a continuum source would have, namely, one source for all determinations rather than needing a separate hollow-cathode lamp for each determination. One very promising means of overcoming this limitation was first proposed by Margoshes (9). This would involve the combination of a monochromator and an interferometer to achieve a system with a spectral bandwidth narrower than the absorption line width. Thus, one would achieve with a continuum source the same sensitivity as with a narrow line source, without the expense of a high resolution monochromator (see Chapter 5, IV. E).

C. Fuels, Oxidants, Burners, and Nebulizers

For optimum results in the analysis of a particular element by flame emission, absorption, or fluorescence, the burner design, sample introduction system, fuel, and oxidant (and their relative flow rates) must be carefully considered. One might also consider the use of nonaqueous solvents or miscible organic solvents, in some cases because of their strong effect in increasing the efficiency of atomization (10). Most burners used in flame spectrometry fall into two general categories: the total consumption, external-mix type and the premixed, chamber type. The former sprays (nebulizes) all of the sample directly into the flame. Flame gases (fuel and oxidant) are fed to the burner separately and mix external to the burner. One of the gases, usually the oxidant, is used to aspirate the sample into the flame. This type of burner is widely used in flame spectrometry, especially for flame emission and atomic fluorescence. Well-known examples of this type of burner are the Beckman and Zeiss burners. The chamber type is the burner most widely used in atomic absorption spectrometry. Here a separate nebulizer (usually pneumatic, as in the Beckman type) sprays the sample solution into a chamber where the larger particles are removed ("condensed" on the chamber walls), and only the finer aerosol particles are carried into the flame. Usually the oxidant is used to aspirate and nebulize the sample. Fuel is also introduced into the chamber, allowing the fuel and oxidant to mix before exiting at the burner head and forming the flame. In atomic absorption, the flame (i.e., the exit aperture of the burner) is often an elongated slot, increasing the absorption path length through the flame. Details of burner construction and the properties of various oxidant-fuel combinations have been described in Chapter 6 of Volume 1 and Chapter 3 of this volume, so they will not be repeated here. Our concern here will be with optimization of these parameters, especially the gas flow rates.

Often one does not have many different burners available to try in an analysis, but if more than one is available, they should be evaluated in terms of sensitivity and detection limit (i.e., signal-to-noise ratio). Selection of the fuel and oxidant will often be determined by the burner design. The most commonly used fuel-oxidant combinations are oxygen–acetylene, air–acetylene, nitrous oxide–acetylene, oxygen–propane, air–propane, oxygen–hydrogen, and air–hydrogen. Selecting the optimum combination is somewhat difficult and usually one must resort to experimentation with various flow ratios or consult texts and the literature for data on the specific problem. Often one can predict what the optimum, or nearly optimum, combination will be from a knowledge of the high-temperature chemistry of the element to be determined and the properties of the resultant flame. Let us look briefly at some examples and make a few generalizations.

In emission flame spectrometry for the determination of easily ionizable elements, such as the alkali metals, one is not often concerned with compound formation but is concerned with the degree of ionization and excitation in the flame. Too high a flame temperature could cause an appreciable fraction of the atoms to be ionized. Therefore, one would usually find that a relatively low-temperature flame would be optimum in terms of sensitivity and signal-to-noise ratio. Cooler flames, like air–propane, air–hydrogen, air–acetylene, or oxygen–propane would probably give better results than the hotter flames. These flames should also work well in atomic absorption and fluorescence for elements that do not show a strong tendency toward compound formation (Zn, Cu, Ag, Cd, In). For emission measurements on elements with appreciable excitation energies and ionization potentials, but no strong compound formation tendencies, one would desire a high-temperature flame, to increase the degree of excitation and, hence, the emission. Flame gas combinations like nitrous oxide–acetylene, oxygen–acetylene, and oxygen–hydrogen should prove to be optimum for elements of this type by flame emission. Elements that show a strong tendency toward compound formation (usually monoxide) require a somewhat different flame; not different in the gases used, but in the flame conditions (flow ratios). Examples of elements of this type include V, Al, Sn, and Ti. In these cases, a high-temperature flame is desired to increase the dissociation of the metal monoxide. Also, a strongly reducing flame atmosphere is desired to chemically reduce the monoxide. For example, aluminum atoms in a normal, stoichiometric flame (e.g., oxygen–acetylene) exist almost entirely as the oxide.

By making the flame very fuel-rich and perhaps using an organic solvent, highly reducing species, such as free carbon and carbon monoxide, are produced in the interior of the flame and these can reduce the metal oxide (see Chapters 6, 11, and 12 of Volume 1). Usually a fuel-rich oxygen–acetylene or nitrous oxide–acetylene flame is optimum. An oxygen–hydrogen fuel-rich flame will not usually be as effective because no very strongly reducing species are formed. However, when organic solvents are employed it can be quite effective since it is a high-temperature flame.

Once one selects a flame gas combination, the most critical parameters to be optimized are the flow rates of each and the flow ratio of the gases (i.e., the relative amounts of each being fed into the flame). For this reason it is important to be able to measure the flow rates of the gases and detect any changes in these rates, as discussed earlier. When optimizing these parameters it is important to keep in mind that the signal-to-noise ratio is as important, if not more so, than the sensitivity. In other words, one is not trying to maximize the signal alone without consideration of the signal-to-noise ratio.

Usually the oxidant is used to aspirate and nebulize the sample solution, so the control of the oxidant flow rate is most important in determining the sample introduction rate. Fuel flow is not usually so critical except under fuel-rich conditions and in emission measurements. Usually a given burner-nebulizer system will have optimum flow rates and ratio of the gases (assuming one of the gases determines the sample introduction rate) for the analysis of a given element by a given technique. These optimum conditions result in the highest signal-to-noise ratio and maximum sensitivity. One procedure frequently used to optimize these parameters is to measure the signal (i.e., sensitivity) and the noise (for the signal-to-noise ratio) under various gas flow conditions and rates. If the oxidant flow rate determines the sample introduction rate, one might set a given oxidant flow rate and vary the oxidant-fuel flow ratio (by adjusting the fuel flow) over a range (e.g., stoichiometric to several-fold fuel-rich) maximizing the signal-to-noise ratio at that oxidant flow. Then the oxidant flow could be changed, and the process repeated. One can quickly find the conditions near the optimum in this manner. Increasing the oxidant flow will, up to a point, increase the sample uptake rate but usually decrease the efficiency of atom production in the flame (10). Usually one would look only on the fuel-rich side of the stoichiometric flame for optimum conditions since the optimum flow ratio will rarely occur on the fuel-lean side. One exception is in analyses by emission where a molecular band (e.g., oxide) rather than an atomic line is employed. In the oxygen–acetylene flame, the reaction is

$$C_2H_2 + 2.5O_2 \rightarrow 2CO_2 + H_2O$$

Thus, the flow rate of oxygen is $2\frac{1}{2}$ times the acetylene flow in the stoichiometric flame (actually slightly less, due to the entrainment of atmospheric oxygen). Fuel-rich flames as high as 1:1 acetylene:oxygen and higher have been employed with the more refractory oxides. As the flow ratio of fuel-to-oxidant is increased, one will find an optimum ratio where the signal-to-noise ratio is a maximum. The signal-to-noise ratio need not necessarily be calculated each time, as it is often obvious in what range of conditions it is a maximum. This is especially true when a recorder is employed as the readout device on the instrument.

IV. Measuring or Recording the Signal

In Chapter 6 of this volume various readout devices for spectrometers are discussed. At present, the two most popular are the meter and the recorder. Deflection meters have certain disadvantages as readout devices. First, it is difficult to get an accurate measurement of the noise magnitude and, hence, difficult to determine the signal-to-noise ratio for the measurement. Likewise, under conditions of high noise, it is difficult to accurately determine the signal level. Second, there is always the possibility that they will be misread and the number written down is in error. Third, any interpolation to be made by the "eyeball" technique is highly subject to operator bias. This can often eliminate one significant figure from the reading. The digital meter readout eliminates the third, and reduces the second disadvantage. It is still as difficult to measure the noise magnitude, if not more so. The recorder overcomes the first disadvantage handily. As for the second, it is possible to misread the chart, but since it is a more-or-less permanent record it can always be rechecked. It, too, can suffer from operator bias, although it is usually much less of a problem due to the larger scale and smaller need for interpolation.

In measuring or recording the signal in an analysis by emission, one must take into account any background emission by the flame itself. It is the line-above-background intensity that is to be determined. Changes in the background intensity may or may not change the emission line intensity. This is one situation where a scanning monochromator-recorder combination permits one to scan over the wavelength region of interest and record directly both the line intensity and the background emission intensity on either side of the line. One should also scan the same region with a blank solution to be certain that the background emission with the blank is the same as with the sample. If so, one can then determine the emission line intensity (above background) from the chart. With a non-scanning, meter-type instrument, the monochromator wavelength setting is "peaked," using a standard solution of the desired element. Samples are then measured, correcting for the flame background emission with the blank solution.

In atomic absorption spectrometry one does not usually scan the monochromator wavelength setting, except to ensure that no extraneous source emission lines are within the spectral bandwidth of the instrument (or to determine their spectral distance from the desired line). One can determine from this the maximum spectral slit width to be employed. Now the monochromator wavelength setting is peaked on the line. The source intensity (flame on, blank being aspirated) is adjusted by whatever means is provided (amplifier gain, photomultiplier voltage) to a full-scale reading (e.g., $100\% \; T$). Then the source is blocked off and the zero adjustment is made (e.g., $0\% \; T$). If a meter readout is employed (reading transmittance, in our example), standard and sample solutions are aspirated and the results written down. The 0 and 100% readings should be checked occasionally to be certain that they have not changed. If a recorder is employed as the readout, the same procedure is followed except that it is not necessary to write down the values immediately since they can be read from the chart later.

In atomic fluorescence spectrometry the procedure for measuring the signal is quite simple. Almost always a recorder is used in taking the measurements. Sample and blank are aspirated, and the relative intensities of each are recorded.

V. General Trouble-Shooting Suggestions

In this section the word "general" must be emphasized. Most instrument instruction manuals have a trouble-shooting section which deals with that specific instrument. This is especially important when trouble with the electronics or optical-mechanical parts of the instrument are encountered. Indeed it would be difficult to generalize about electronic problems without knowing the particular details of the circuits used. (See the discussion on fault diagnosis, Chapter 1, V. B.)

When dealing with solutions with high concentrations of total solids, clogging of the burner or nebulizer becomes a problem, especially with sprayer-burner combinations. Even a slot burner may suffer from partial clogging of the nebulizer or burner opening. Concentration of solute above which this problem becomes serious is difficult to define; often clogging occurs for concentrations 5000 μg/ml and greater. Deposits building up around the tip of the nebulizer impede the solution aspiration rate, affect the flow of aspirating gas, and, consequently, markedly alter the measurements. When external light sources are used in atomic absorption or atomic fluorescence, scattering of source radiation by solid particles in the flame can also be a problem with solutions of high solid concentration (Chapter 10 of Volume 1). Usually clogging and scattering are first noted by a decrease in sensitivity; other indications include a change in the aspirating gas-flow rate, a change in the solution uptake, and a change in the appearance or sound of the flame. Frequent rinsing of the nebulizer-burner system, preferably after each sample, with solvent and limiting the total aspirating time of concentrated solutions before rinsing are remedial actions to overcome clogging. Scattering requires correction through use of a nearby nonabsorbing source line.

When attempting to measure a major constituent in the same solution used for measuring the minor constituents there usually is such a high concentration of analyte that very little source radiation in atomic absorption is transmitted at the optimum wavelength. One remedy is to select a less strongly absorbing line, another is to decrease the absorption path either by changing from a 10-cm slot burner to a 5-cm slot burner or by rotating the slot burner until it is at a right angle to the optical axis.

Long-term changes in sensitivity are often noted when using commercial acetylene as the fuel. This can sometimes be related to the tank pressure. A tank of commercial acetylene might contain initially a pressure of 250–300 psi (17.5 − 21.0 kg/cm^2). By the time the cylinder pressure drops to something below 50 psi ($<$ 3.5 kg/cm^2) the sensitivity might be only half of what it was initially. This is due to the fact that below 50 psi an appreciable amount of acetone vapor is coming out with the acetylene. Acetone is used in the tank to dissolve the acetylene and allows it to be pressurized above 15 psi (1.0 kg/cm^2). Incidentally, acetylene gas should never be released at pressures greater than 15 psi. For the reason mentioned above, acetylene tanks should be changed before they reach a pressure of 50 psi, preferably at 75 psi (5.2 kg/cm^2) or higher. Copper tubing should not be used to transport acetylene, because of the possibility of acetylide formation.

Another problem which may be encountered is detector fatigue, especially with photomultipliers. The causes, symptoms, and cures for this malady have been described in Chapter 6 of this volume.

When making measurements at fixed monochromator wavelength settings, the peak wavelength setting may drift slightly with temperature changes. This often occurs most noticeably when the flame is first lit, due to its heating effect on the instrument module. After about 15–30 min equilibrium is attained and the effect should disappear. It can also occur if the laboratory has poor (or none at all) thermal regulation, such as air conditioning or heating systems with insufficiently sensitive thermostats. In this case, the wavelength setting drift will persist, and one must check frequently to be sure that it is maximized.

Short-term and/or long-term instability in the source (absorption and fluorescence) can result in noise and/or drift, respectively, in the signal. If due to the source lamp itself, little can be done except to replace the source. If the instability is due to the source power supply, the power supply should be repaired, modified, or replaced. Often this instability is due to fluctuations in the line voltage (house mains). Most power supplies have some means of input voltage regulation to eliminate this effect. Many do not, or have inadequate regulations, and a constant voltage regulator is required.

When using a recorder one might notice that the baseline does not immediately return to its previous value after a reading is made. It also might be noticed that it does not give the same reading of a signal when approached from different directions. These problems could be due to the "deadband" of the recorder being too large and can usually be remedied by increasing the gain of the recorder's amplifier. Conversely, if the trace tends to "overshoot" on a reading, or if there is a tendency for the pen to oscillate, the gain is too high.

Fluctuations in the gas flow rate, often due to pressure changes, can be troublesome. Single-stage gas pressure regulators will often change in their delivery pressure as the tank pressure changes. This is much less of a problem with 2-stage regulators. Too small a flow regulator can also cause problems in the form of a cooling effect due to the pressure drop. This is especially troublesome with nitrous oxide gas. Frequently it is necessary to warm the regulator slightly with heating tape. These temperature changes alter the regulator adjustment, altering the pressure and, consequently, the flow rate of the gas. Another cooling effect can occur with chamber-type burners. This is due primarily to solvent evaporation in the chamber. As the chamber temperature drops, the efficiency of the sample introduction system also drops (i.e., less sample reaches the flame), causing a drift. One means of preventing this is to constantly aspirate solvent when not aspirating solution. This will allow the chamber temperature to stabilize and will also reduce the clogging problems mentioned earlier.

It will also reduce another problem, namely, carryover. Carryover (memory) can occur in some cases with chamber-type burners and fairly high concentrations. Here, it takes a finite time for the chamber to completely rid itself of the aerosol of the previous sample. Carryover can be detected when aspirating solvent after the sample, or by running the first set of samples in increasing concentration and repeating by running them in reverse order. With strongly acidic solutions, corrosion can be a problem. Spray chambers must be equipped with proper gaskets to resist acids and also organic solvents.

In atomic absorption spectrometry, any unabsorbed source radiation reaching the detector will reduce the sensitivity of the analysis. This would occur if the spectral bandpass of the monochromator was too wide, so that a nearby line (unabsorbed) is also included. For example, suppose the line of a certain metal emitted from a hollow-cathode lamp has another line due to the fill gas (e.g., neon) 5 Å away from it. If the monochromator wavelength setting is peaked on the metal line (narrow slits) and the spectral bandpass adjusted (by widening the slits) to a value of 7 Å, the fill gas line will not reach the detector. Only wavelengths ± 3.5 Å on either side of the metal line will be passed, excluding the fill gas line. If the spectral bandpass is increased to 12 Å (± 6 Å on either side of the metal line), the fill gas line only 5 Å away will be passed and reach the detector. It will not be absorbed by metal atoms in the flame, but will contribute to the overall measured signal. In measuring the absorbance ($\log I_0/I$) in the first case, $I_0 = I_{\text{metal}}$ and $I = I_{\text{metal}} - I_{\text{absorbed}}$. In the second case, $I_0 = I_{\text{metal}} + I_{\text{neon}}$ and $I = I_{\text{metal}} + I_{\text{neon}} - I_{\text{absorbed}}$. Thus, the measured intensity with sample (I) in the second case will be greater than it should be by an amount equal to I_{neon}, and the absorbance will consequently be less, lowering the sensitivity.

REFERENCES

1. J. M. Mansfield and J. D. Winefordner, *Anal. Chim. Acta*, **40**, 357 (1968).
2. C. H. Corliss and W. R. Bozman, *Experimental Transition Probabilities for Spectral Lines of Seventy Elements*, NBS Monograph 53, U.S. Govt. Printing Office, Washington, D.C., 1962.
3. P. T. Gilbert, Jr., in *The Encyclopedia of Spectroscopy* (G. L. Clark, ed.), Reinhold, New York, 1960, pp. 350–355; *Bulletin 753-A*, Beckman Instruments, Inc., Fullerton, Calif., 1961.
4. K. M. Cellier and H. C. T. Stace, *Appl. Spectry.*, **20**, 26 (1966).
5. A. Strasheim and G. J. Wessels, *Appl. Spectry.*, **17**, 65 (1963).
6. J. D. Winefordner, *Appl. Spectry.*, **17**, 109 (1963).
7. M. L. Parsons and J. D. Winefordner, *Appl. Spectry.*, **21**, 368 (1967).
8. J. D. Winefordner and T. J. Vickers, *Anal. Chem.*, **36**, 1939 (1964).
9. M. Margoshes, private communication.
10. J. D. Winefordner, J. M. Mansfield, and T. J. Vickers, *Anal. Chem.*, **35**, 1607 (1963).

8 Atomic Fluorescence Spectrometry

Augusta Syty

DEPARTMENT OF CHEMISTRY
INDIANA UNIVERSITY OF PENNSYLVANIA
INDIANA, PENNSYLVANIA

I. Introduction

Atomic fluorescence flame spectrometry is an emission method of analysis which is based upon radiational activation of the atomic vapor in the flame. Molecular fluorescence as an analytical tool has been known for a long time, and the very phenomenon of fluorescence emitted by atoms has also been observed since the end of the last century. But the analytical use of the fluorescence emitted by atoms is only now being developed.

A. HISTORICAL

Going back to the early work on atomic fluorescence, one finds that possibly the first observation of the phenomenon of atomic fluorescence has been the work of Wood around the turn of the century (*1*). He observed the fluorescent *D* lines of sodium. Work with the excitation of Na vapor has been attempted before him, but the resonance lines of sodium were never detected. Typically, these early workers used a glass bulb in which they produced some sodium vapor by heating the sodium metal; they illuminated the bulb with sunlight or with the flame fed with a sodium solution and expected to see the yellow fluorescence of the *D* lines. The reason for their failure to see any fluorescence was that, at the high vapor pressure of sodium employed during those experiments, the vapor was mainly in the form of Na_2 molecules, and the fluorescence spectrum consisted of bands of this molecule rather than only lines. Wood was successful because he generated a low pressure of sodium vapor in an evacuated test tube. When he illuminated it with a flame containing NaCl vapor, a yellow fluorescence was apparent in the test tube, which consisted of the *D* lines.

Atomic fluorescence of several other elements (*1*) was also observed in the early 1900's, but atomic fluorescence found its main use in astrophysical work, for the study of the composition of atmospheres of various stars and suns, and not in analytical applications. All of the early workers confined the atomic vapor in glass or quartz cells, which is much less convenient than the present use of a flame as a sample cell.

In flames, atomic fluorescence was first reported by Nichols and Howes (*2*) in 1924. They detected weak fluorescence of Na, Li, Ca, Sr, and Ba from high concentrations of these atoms in the flame. Many years later, Robinson (*3*) detected a weak atomic fluorescence signal for Mg from an oxygen–hydrogen flame irradiated' by a Mg hollow-cathode lamp. He did not, however, detect any similar emissions from sodium or nickel. None of the above workers were interested in the phenomenon from the point of view of its analytical usefulness, and it remained for the investigators of the last few years to develop the technique into a very promising trace method of analysis.

In 1962, Alkemade (*4*) reported the use of atomic fluorescence for determination of excitation processes in the flames and for evaluation of quantum yields for the excited atoms. He pointed out the analytical possibilities of the method. Atomic fluorescence as an analytical tool was introduced in 1964 by Winefordner and his co-workers (5–7). Since then, a growing number of elements have been analyzed, the theory of the method has been developed, and improvements and modifications in instrumentation, primarily in the excitation sources, are being reported continuously (8).

B. Basic Apparatus

The typical experimental arrangement which may be used for atomic fluorescence work is illustrated by the block diagram in Fig. 1. The atomic vapor generated in the flame cell is irradiated. The atoms absorb their resonance lines and rise to higher excitation states. In the fluorescence spectrum obtained when excited atoms return to ground state, the wavelengths of the emitted lines are characteristic of the element and the

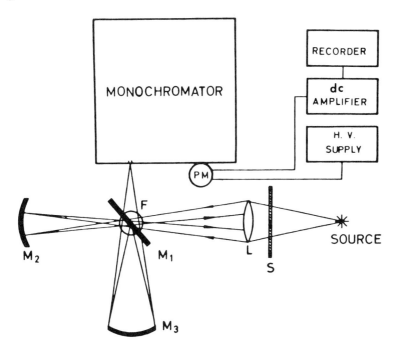

Fig. 1. Block diagram of an atomic fluorescence arrangement: M_1 is a rotatable plane mirror to peak the exciting radiation from the source; M_2 and M_3 are concave mirrors; F is the flame; S is a shutter.

intensity of the lines is characteristic of its concentration. Fluorescence is emitted in all directions equally, and some of this radiation falls on the detector which views the flame at right angles to the excitation beam. The amplified photomultiplier signal is displayed on a meter or a recorder. Some parts of instrumentation will be discussed later, but it can be clearly seen that atomic fluorescence spectrometry combines some aspects of flame emission spectrometry with some characteristics of atomic absorption spectrometry and, consequently, many parts of instrumentation will resemble closely those employed in the other two flame methods.

II. Theory (9)

Atomic fluorescence may be divided into three main types (also see Section II.C) which are defined as follows. *Resonance fluorescence* is the process by which an atom, excited by a given spectral line, returns to the ground energy level by emitting the same spectral line. *Stepwise-line fluorescence* involves the excitation of an atom to an energy level above the first excited state, which is followed by a nonradiational deactivation to the lower excited state, and which, in turn, is followed by a radiational return to ground state. *Direct-line fluorescence* also involves the excitation of an atom to an energy level above the first excited state, but this is followed by a radiational deactivation to the lower excited state, followed

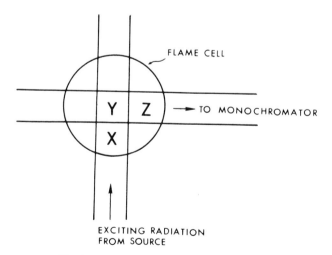

Fig. 2. Horizontal cross section of the flame.

next by a nonradiational deactivation to the ground state. These three types of atomic fluorescence can be illustrated schematically as follows:

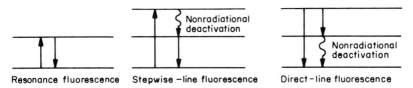

Resonance fluorescence   Stepwise –line fluorescence   Direct – line fluorescence

Expressions for the intensity of fluorescence can be derived based on the classical theory of radiative transfer. In order to incorporate the effect of the geometry of the flame cell into the quantitative expressions, the flame cell can be represented as in Fig. 2, with the three analytically important regions of the flame designated X, Y, and Z. Regions X and Y are illuminated by the exciting radiation. Regions Y and Z are viewed by the monochromator. Region Y is the analytical part of the flame which is excited and whose emission is measured.

A. INTENSITY OF THE ATOMIC FLUORESCENCE LINE

If atomic fluorescence is measured under ideal experimental conditions such that no incident exciting radiation finds its way into the monochromator and thermal emission and flame background emission are either not present or eliminated, the integrated signal intensity measured at the atomic fluorescence line is given by

$$I_F = I_A \left(\frac{\Phi}{4\pi}\right) B_z \tag{1}$$

In this equation I_A is the intensity absorbed by the resonance line, Φ stands for fluorescence efficiency, and equals the ratio of the power emitted per unit time as fluorescence to the excitation power absorbed per unit time. It is related to the quantum efficiency ϕ by the equation $I = (v_f/v_a)\phi$, where v_f and v_a are the frequencies of the fluorescence and the absorbed photons, respectively. 4π is the number of steradians in a sphere, and B_z is a factor which accounts for the decrease in fluorescence intensity due to absorption by particles in region Z of the flame, which is not illuminated by the exciting radiation but which has to be traversed by the fluorescence radiation before it reaches the monochromator.

A more explicit equation for the intensity absorbed by the atoms that fluoresce is given by

$$I_A = I_0 B_x \Omega A_t \tag{2}$$

where B_x is a factor which accounts for the decrease in the intensity of the exciting radiation by absorbing atoms in the region X of the flame, which is in the path of the exciting beam but not in the line of view of the monochromator. Ω is the solid angle of the exciting radiation incident upon the flame cell. A_t is the total absorption factor which equals the fraction of incident radiation absorbed by atoms in the region Y of the flame; and I_0 is the intensity of the exciting radiation, for a source which emits a wide line or a continuum. If the source emits a very narrow line, I_0 must be replaced by

$$I\left(\frac{2(\ln 2)^{1/2}}{\pi^{1/2}}\right)(\Delta\nu)$$

where I stands for the intensity of the source line with a Gaussian shape, and $\Delta\nu$ represents the source line half-width.

To evaluate the total absorption factor A_t, one integrates over the width of the absorption line, provided a continuous or a wide line source is used. For a very narrow source line, the integration would be over the source line half-width.

The factors B_x and B_z approach unity as the solid angle over which the excitation source irradiates the flame is increased to encompass the entire flame cell and the solid angle over which the monochromator views the flame is increased to include the entire flame cell. For the expressions used in the evaluation of factors B_x and B_z the reader is referred to the original papers.

Combination of Eqs. (1) and (2) yields the general expression for the intensity of a resonance fluorescence line:

$$I_F = I_0 B_x \frac{\Omega}{4\pi} A_t \Phi B_z \qquad (3)$$

To make Eq. (3) applicable to stepwise-line and direct-line fluorescence, all absorption lines (besides the resonance line) that may excite the fluorescent energy level must also be considered. If there are j such absorption lines, the expression for the intensity of fluorescence becomes

$$I_F = \left(\frac{\Omega}{4\pi}\right) B_z B_y \Phi \sum_j (I_0)_j (B_x)_j (A_t)_j \qquad (4)$$

The additional factor B_y stands for the fraction of fluorescence radiation (caused by absorption of lines other than resonance) which travels through the analytical region Y of the flame without suffering reabsorption by other atoms of the same kind. Reabsorption of the resonance fluorescence alone in the region Y is inherent in the total absorption factor A_t. Thus, for resonance fluorescence, B_y equals unity, and Eq. (4) reduces to Eq. (3).

B. Dependence of Atomic Fluorescence Intensity upon Concentration and upon the Source Line Width

As can be seen from the above equations, the intensity of fluorescence I_F varies directly with the magnitude of the total absorption factor A_t. The total absorption factor A_t is clearly a function of the concentration of the absorbing atoms in the flame. The nature of this dependence, in turn, depends upon the width of the atomic absorption line relative to the width of the exciting line. Since the analyst is usually interested in how the intensity of fluorescence is related to concentration, it is of interest to consider how A_t, and therefore I_F, varies as a function of atomic concentration at low and high values of concentration for the two extreme cases: (a) the absorption line much narrower than the source emission line and (b) the absorption line much wider than the source emission line.

The equations describing A_t in these extreme cases have been derived and reveal that at low concentrations the intensity of fluorescence varies directly with concentration for both the narrow and the wide source lines. At high concentrations, the intensity of fluorescence varies directly with the square root of concentration for the wider source line and is independent of concentration for the narrow source line. These results are illustrated qualitatively in Fig. 3.

In the top two diagrams in Fig. 3, which apply to the continuous source or a wide source line, the area under each absorption curve represents the total absorption factor A_t. One observes that, for low concentrations, the area under the curve varies directly with concentration. For high concentrations, the area within the profile of the line source and therefore A_t does change with concentration but not as rapidly. Actually, A_t changes with the square root of concentration.

In the bottom diagrams in Fig. 3, which apply to a narrow line source, two types of curves are superimposed. The shaded area within the curve represents the emission line coming from the excitation source. The rest of the lines again represent the absorption lines. The energy absorbed, or the absorption factor A_t in this case, is not the total area under the absorption curve, but only that portion that is also shaded. One observes that for low concentrations A_t varies directly with concentration, but for high

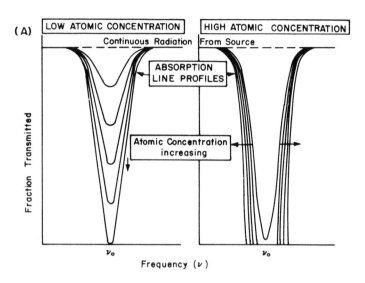

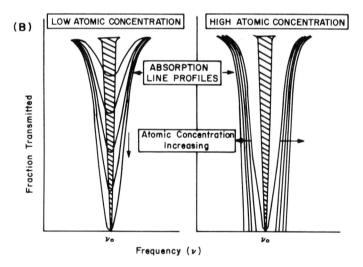

Fig. 3. Variation of total absorption of radiation with atomic concentration from continuous source (A) and line source (B). [Reprinted from (9) by courtesy of Pergamon Press.]

concentrations, as soon as the concentration is high enough to absorb all of the emitted excitation energy, any further increase in concentration does not result in any increase in absorption. Thus, A_t becomes independent of the concentration of the atoms in the flame.

From Fig. 3 it is clear that the calibration curves of log I_F versus log N_0 (where N_0 is the concentration of the atoms of interest in the flame) are expected to have the general shapes indicated in Fig. 4.

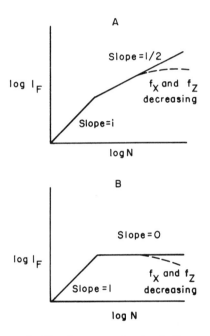

Fig. 4. General forms of calibration curves in atomic fluorescence: (A) source line half-width greater than absorption line half-width; (B) source line half-width smaller than absorption line half-width. [Reprinted from (9) by courtesy of Pergamon Press.]

As was pointed out, the value of the factors B_x and B_z would be unity if the entire flame cell were illuminated by the exciting radiation and if the entire fluorescence radiation were measured. Since in practice only a fraction of the flame cell is illuminated and only a fraction of the fluorescence radiation is measured, these factors are considerably less than unity and approach zero for high concentrations of the absorbing atoms. Thus, the calibration curves are expected to exhibit pronounced curvature toward the concentration axes at high concentrations.

Although N_0 stands for the total concentration of the atoms of interest in the flame, the concentration of an element in a sample solution, C, may be substituted on the horizontal axis of the theoretical calibration curves. A simple empirical equation that relates these concentration terms is

$$C = \frac{KN_0}{\xi\varepsilon} \tag{5}$$

where K is a constant whose magnitude reflects the nature of the flame, the flow rates of fuel and oxidant, and the rate of sample consumption; ξ is the efficiency of nebulization and vaporization, and ε is the efficiency of atomization (10).

An expression has been derived (10) for calculation of the limiting atomic concentration detectable by atomic fluorescence using the signal-to-noise theory. The limiting concentration is expressed as a function of various parameters like the intensity and noise of the excitation source, flame background intensity and noise, power efficiency for fluorescence, monochromator slit width, and amplifier type. However, from the practical point of view, it is much more convenient to determine the detection limit experimentally. Also a complex equation has been derived (11) which expresses the fluorescence power efficiency as a function of numerous instrumental parameters.

C. SENSITIZED FLUORESCENCE

Sensitized fluorescence (1) constitutes the fourth variety of atomic fluorescence in addition to resonance, stepwise-line, and direct-line types of fluorescence which have already been mentioned. In sensitized fluorescence, the atom of interest fluoresces after becoming activated by a collision with a foreign atom which had previously been excited by absorption of resonance radiation. For example, consider a mixture of atoms A and atoms B. The mixture is irradiated by light of frequency ν. Atoms A absorb it. If the number of atoms B is high enough, so that the time between collisions of A with B is of the same order of magnitude as the mean life of the excited state of A, then energy will be transmitted to B by collision and some fluorescence of the frequency ν' will also be seen, as illustrated in Fig. 5.

The energy difference ΔE between the two excited states will show up as the relative kinetic energy of A and B atoms. If ΔE is large, the atom B

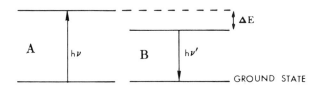

Fig. 5. Schematic diagram of the process of sensitized fluorescence.

will acquire a considerable energy, especially if it is a light atom. This fact is confirmed by the Doppler broadening of the sensitized fluorescence line of frequency ν'.

The classical example of sensitized atomic fluorescence involves illumination of a mixture of Hg and Tl atomic vapors by the 2537-Å Hg line, which results in appreciable fluorescence of the Tl lines at 5350 and 3776 Å.

III. Instrumental Parameters

The schematic diagram of the typical experimental setup required for atomic fluorescence is illustrated in Fig. 1. Commercially available flame photometers and atomic absorption instruments can easily be modified for fluorescence work. The main modification is the introduction of a source lamp to irradiate the flame at a 90° angle to the direction in which the monochromator views the flame.

A. Sources

1. *Types of Excitation Sources*

According to Eq. (3) the intensity of atomic fluorescence, I_F, increases directly with the intensity of the exciting radiation, I_0, coming from the source. Thus, it is clear that intensity is the main criterion in selecting an excitation source which is to yield the best possible sensitivity for the atomic fluorescence technique. In addition, of course, the source should be stable and preferably have a short warm-up period.

One of the first sources employed in atomic fluorescence spectrometry was an ordinary hollow-cathode lamp. The excited fluorescence was very weak (*12*), and this is not surprising since hollow-cathode lamps are designed with a different set of desirable characteristics in mind. For work in atomic absorption spectrometry an excitation source of moderate intensity and with very narrow emission lines is needed. A very intense source would not improve the atomic absorption measurements but increase the susceptibility to interference by atomic fluorescence; wide source emission lines would decrease sensitivity and impart more curvature to the calibration curves. In atomic fluorescence, on the other hand, the intensity of the excitation source is of primary importance, and it can not, in a sense, become "too" intense, since none of the exciting beams fall on the detector. Furthermore, the width of the source line is of no particular importance, provided the emission is intense at the wavelength of absorption. As a result of their low intensities, ordinary hollow-cathode lamps are not used in atomic fluorescence work.

High intensity hollow-cathode lamps are a significant improvement over the conventional kind (13, 14). In spite of the fact that they are relatively expensive and require two power supplies, they are quite suitable for many elements and yield fluorescence detection limits at least as good as those obtained with electrodeless discharge tubes.

Demountable water-cooled hollow-cathode lamps have also been used (15, 16). These lamps seem to offer satisfactory intensity, but the process of replacing the cathode and evacuating the envelope and introducing a low pressure of inert gas may be inconvenient in routine analysis. The demountable hollow-cathode lamp proved too unstable for Mg, but good responses were obtained for Cu, Ag, Au, Pb, Bi, Tl, Ni, and Ga (15).

Other excitation sources which were employed very early in atomic fluorescence work are Osram and Philips spectral discharge lamps (6, 7, 17–20). The emitted lines are wide and intense, but, unfortunately, they are very self-reversed in many cases and are available for relatively few elements. Spectral discharge lamps for Cd, Zn, and Tl give the best fluorescence results.

The continuous spectra of 150-W and 450-W xenon arcs may be used to excite atomic fluorescence (20–22). The obvious advantage of a continuous source is the elimination of the need for an individual source for each element. Continuous sources have been used in atomic absorption work, but their applicability was limited by the need for high-quality monochromators with very narrow spectral bandwidths. In atomic fluorescence work employing a continuous source a much less expensive monochromator suffices, since it is only needed to separate the fluorescence line of interest from other fluorescence lines and from the thermal flame emission background. The relative intensity of the Xe arc at the resonance wavelengths of various elements is, however, much lower than that available from the line sources and falls off especially below $\sim$2500 Å, where many useful lines are located.

Electrodeless discharge tubes are the best excitation sources currently used in atomic fluorescence work (23–27). These tubes consist of a glass or quartz cell containing a low pressure of an inert filler gas and a small amount of a metal or a metal salt. These lamps are operated at radio or microwave frequencies, and have the advantage of being relatively easy to manufacture in the laboratory. They have been prepared for many

TABLE 1

DETECTION LIMITS ATTAINABLE WITH DIFFERENT
EXCITATION SOURCES

Source	Atomic fluorescence detection limit, $\mu g/ml$[a]					
	Zn	Co	Ag	Pb	Ni	Cd
Xe arc, 150 W	0.6 (21)	1.0 (20)	0.08 (21)	7.5 (21)	3 (28)	0.08 (21)
Xe arc, 450 W	0.03 (22)	0.5 (22)	0.001 (22)		1.0 (29)	
Spectral discharge lamp	0.0001 (30)					0.0002 (30)
Demountable hollow cathode		0.5 (28)	0.001 (30)	1.0 (30)	0.5 (28)	
High intensity hollow cathode		0.01 (31)	0.0017 (13)	0.02 (32)	0.003 (31)	
Electrodeless discharge lamp	0.00004 (25)	0.005 (50)	0.0001 (25)	0.5 (25)	0.04 (25)	0.000001 (25)

[a] References are given in parenthesis.

elements, emit very intense and stable spectra of narrow lines, essentially free from self-reversal, and appear to have a long life. The parameters of the electrodeless discharge lamps which directly affect the intensity of atomic fluorescence are the diameter of the lamp, the pressure of the filler gas, and whether the element of interest is added as a metal or as a metal salt. Also, the intensity of the source is independent of the amount of the metal or its salt which is introduced into the lamp; however, copious amount of material will block radiation by covering a portion of the lamp wall (26).

A general idea of the relative suitability of various common sources may be obtained from a glance at a few detection limits listed in Table 1.

It has to be pointed out that different authors use different definitions of the detection limit and, for this reason, if adjusted to a consistent scale, the numbers listed in the table may easily vary by a factor of 2 or 3.

It is possible to excite a fluorescence line of an element with an overlapping line of a different element emitted by the excitation source. Mercury exhibits a very line-rich emission spectrum, and a Philips Hg 90-W high-pressure discharge lamp is found effective in exciting fluorescence of Fe, Mn, Ni, Cr, Tl, Cu, and Mg (33). The fluorescence

TABLE 2

ATOMIC FLUORESCENCE OF SEVERAL ELEMENTS
EXCITED BY THE MERCURY DISCHARGE LAMP

Element	Fluorescence line, Å	Hg line, Å	Detection limit,[a] μg/ml
Fe	2483.27	2482.72	1
Tl	3775.72	3776.26	0.3
Cr	3593.49	3593.48	5
Mg	2852.13	2852.42	0.5
Mn	2794.82	Continuum	0.5
Ni	2320.03	Continuum	3
Cu	3247.54	Continuum	0.1

[a] Detection limit is here defined as that concentration that yields a signal twice the noise.

thus excited is quite intense for the lines for which the emission and absorption lines overlap significantly, and the Hg lamp continuum is effective in exciting fluorescence of some elements in the absence of any overlap of absorption with emission lines. Some results obtained with the Hg lamp as excitation source are given in Table 2 (33).

Although detection limits obtained with this excitation source do not rival those attained by electrodeless discharge tubes, the high intensity Hg lamp is an interesting possibility and seems to be at least as useful as a Xe arc continuum.

2. Modulation

In some atomic fluorescence work the excitation source and the detection system are modulated. In a dc system, the detector responds to the background radiation of the flame and to the thermal emission of elements at the wavelength of interest along with the desired atomic fluorescence line. Of course, the flame background emission and the thermal emission are extremely low with most flames commonly used in atomic fluorescence, and, if necessary, these emissions can be corrected for by subtracting from the signal the intensity of a blank determined after closing off the excitation source. However, since all conditions which increase the background-to-signal ratio result in relatively more noisy determinations, modulation is often found to improve the limits of detection significantly. In a modulated system, the source emits a pulsed excitation beam as a result of which the excited atomic fluorescence is also a pulsed signal. Thus, a detector tuned to the modulation frequency rejects all background and thermal emissions from the flame which are continuous, and registers only fluorescence signals. Modulation is clearly advisable with high background flames like oxygen–acetylene or nitrous oxide–acetylene, which are sometimes preferred for the more refractory elements.

3. Selection of the Atomic Fluorescence Line

Not all lines emitted by an excitation source result in a fluorescence line. For example, a high-intensity Ag hollow-cathode lamp emits lines at 2061, 2070, 3281, 3383, 4055, 4211, 5209, and 5466 Å, but only the resonance lines at 3281 and at 3383 Å give detectable fluorescence (13). The best line of an element in atomic fluorescence spectrometry is the line for which, assuming no interferences, the combination of source intensity, absorptivity, and detector response is the best. Often, this is not the line which is the most sensitive in atomic absorption. The sensitivity of atomic absorption is almost completely independent of the absolute intensity of the line emitted by the excitation source, since the measured signal is the ratio of intensities of the excitation beam before and after absorption. By contrast, atomic fluorescence is a strictly single-beam system, where the magnitude of the signal is directly dependent upon the intensity of the excitation source and upon the sensitivity of the detector at the particular wavelength. For example, the relative intensity of the two pronounced Ag fluorescence lines, 3281 and 3383 Å, varies from 3:2 to 2:1 going from a 150-W Xe arc continuous source (20) to a high-intensity hollow-cathode lamp, using similar detectors (13).

The choice of the best atomic fluorescence line is usually made on the basis of the experimentally measured intensies, background, and noise for a series of lines. Although, it offers best accuracy, this trial-and-error approach may be somewhat time-consuming since elements emit fluorescence spectra of varying complexities. Table 3 illustrates the fluorescence spectrum of arsenic (24). The alternate approach is the theoretical evaluation of the relative fluorescence intensities resulting from different combinations of transitions which may occur in a given atom returning to ground state, based upon oscillator strengths and other physical constants associated with the various transitions and upon the various experimental and instrumental variables (35). Such calculations narrow down the

TABLE 3

RELATIVE INTENSITIES OF ATOMIC FLUORESCENCE LINES
OF ARSENIC

| Wavelength, Å | Relative signal intensity | |
	Air–C_2H_2 flame	N_2 (entrained air)–H_2 flame
1890	26	37
1937	40	40
1972	40	43
2288	58	4
2350	168	144
2380	33	58
2437	9	13
2457	45	53
2493	44	61
2745	Flame background too high	0
2780	Flame background too high	0
2861	Flame background too high	40
2899	Flame background too high	6
3033	Flame background too high	13
3120	Flame background too high	Too close to OH band

possibilities to a few lines which are most likely to exhibit greatest fluorescence intensities under the given conditions. The obvious limitation of this approach is that it fails to take into account any quenching or interference effects and may thus result in an improper choice of the analytical line.

In the presence of spectral or other interferences, the choice of the analytical line is further limited. For example, in an unmodulated system, thermally excited emission becomes relatively more significant as lines further away from the uv are considered. Table 4 compares the relative intensities of thermal and fluorescence emission signals of thallium (34).

TABLE 4

BACKGROUND, THERMAL EMISSION, AND ATOMIC FLUORESCENCE OF
THALLIUM AT THREE DIFFERENT WAVELENGTHS[a]

Wavelength, Å	Flame background and source scatter	Thermal emission	Atomic fluorescence	Ratio of atomic fluorescence to thermal emission
2768	30	0	8	∞
3776	20	7	44	6.2
5351	25	2	16	8.0

[a] In arbitrary units.

B. FLAMES AND BURNERS

In choosing the composition of the flame and the configuration of the burner and nebulizer to be used for analysis by atomic fluorescence, attention is given to the following points. As usual in all flame methods, the rate of aspiration and the efficiency of nebulization and of atomization are considered, as well as the intensity and the noise of the flame background. In addition, in atomic fluorescence work, the quenching of fluorescence by flame gases and the scattering of the exciting radiation by droplets of solvent and by unvaporized particles of solute play an important role.

1. *Fuels and Oxidants*

The relatively cool hydrogen flames are best applicable to the fluorescence studies of elements such as Cd, Fe, Pb, Tl, or Zn which atomize readily. Hydrogen flames give a low background, little noise, and yield better sensitivities for these elements than the hotter acetylene-containing flames (36). Air–propane is another relatively cool flame which has been found useful for easily atomized elements (13, 20).

The early work in atomic fluorescence was mostly done with an oxygen–hydrogen flame. However, it was soon discovered that some elements gave improved signals with a decreasing flow rate of oxygen and gave best results when the hydrogen flame was maintained by only the entrained air. Ellis and Demers (22) observed improved detection limits for Zn and Ag and especially for Mg and Co. For the latter two elements, the detection limits improved from 1.0 to 0.01 $\mu g/ml$ and from 20 to 1.0 $\mu g/ml$, respectively, when a (entrained air)–hydrogen flame was substituted for oxygen–hydrogen. With elements like Co and Mg, which form oxides, the enhancement is probably due to the more strongly reducing character of the new flame. For the other elements, the enhancement may be due to the decreased possibility of quenching by other flame species.

Veillon et al. (21) observed also that the fluorescence signal of Mg increased when argon was mixed with oxygen of the oxygen–hydrogen flame. The best signal intensity was obtained with pure argon and only sufficient entrained air to maintain combustion. The Mg fluorescence intensity was improved by a factor of 10, and that of Zn, Cd, Tl, and Pb was also increased, although improvement in the detection limits was relatively less pronounced due to an increase in noise in the argon–hydrogen flame. This flame is very similar to an (entrained air)–hydrogen. The signal improvement may be due to decreased quenching of excited atoms by collisions with molecules in the flame and by decreased probability of compound formation in the flame rich in an inert gas. Nitrogen (entrained air)–hydrogen is another similar cool and reducing flame which has been used in atomic fluorescence (24).

The (entrained air)–hydrogen and argon–(entrained air)–hydrogen flames exhibit lower temperatures than the conventional oxygen–hydrogen flame, and as a result more noise is sometimes found due to the less complete vaporization, which cancels out some of the improvement in signal intensity. Smith et al. (37) have determined the temperature profiles of turbulent hydrogen flames used in atomic fluorescence spectrometry and reported the following maximum temperatures:

Argon (entrained air)–hydrogen	2122°K
(Entrained air)–hydrogen	2182°K
Air–hydrogen	2287°K

Apparently the decrease in temperature is not significant enough to cancel out the increased signal intensity from these more reducing flames.

As will be shown, many serious chemical interferences are sometimes encountered in the cooler flames. The use of acetylene-containing flames eliminates these interferences and extends the range of atomic fluorescence spectrometry to some of the very refractory elements. Germanium, for instance, whose chemiluminescence allows only its thermal emission measurement, produces a fluorescence signal at 2652 Å in an oxygen–nitrogen–acetylene flame (38). The function of nitrogen is the generation of a high population of the reducing CN radicals, which help the dissociation of the oxides of germanium. Although acetylene flames assure better vaporization of sample spray and better dissociation of compounds, they generate an intense background which results in flame flicker noise and reduced sensitivity even in a modulated system and at high observation heights. For this reason, acetylene flames are used only rarely.

The nitrous oxide flames employed in atomic absorption work have also found application in atomic fluorescence spectrometry (14, 39). Bratzel et al. (39) tested the nitrous oxide–hydrogen and the nitrous oxide–acetylene flames on several elements. They found the nitrous oxide–hydrogen flame to be the most generally useful of hydrogen flames employed in atomic fluorescence. This flame has a low background, little noise, and a rather uniform temperature over a large part of the flame. The turbulent nitrous oxide–hydrogen flame has a relatively lower rise velocity than other turbulent hydrogen flames, and, as a result, the sample remains in the flame a longer time, allowing a more efficient nebulization. The maximum temperature of a premixed nitrous oxide–hydrogen flame is 2630°K compared to 2250°K for a premixed air–hydrogen flame (37). This makes the flame useful for the relatively less volatile elements; however, in the presence of serious chemical interference, like the presence of significant amounts of silicates for many elements, the hotter and noisier nitrous oxide–acetylene flame is better than nitrous oxide–hydrogen flame.

2. Turbulent and Premixed Flames

Both the turbulent flames from total consumption nebulizer burners, and the premixed flames obtained with spray-chamber burners are used in atomic fluorescence work, although each configuration has some unfortunate limitations. The total-consumption burner is characterized by efficient sample aspiration, but rather inefficient nebulization. This leads to decreased signal intensity, increased background noise, and increased scattering of the exciting radiation. The spray-chamber type burner has a relatively low sample aspiration rate and thus decreased sensitivity, but it generates less noise and causes less scattering by droplets in the flame (40).

Bratzel et al. (40) have compared the relative merits of premixed and turbulent air–hydrogen flames in atomic fluorescence spectrometry and concluded that, although the differences in results are not great, the turbulent flames in general are preferable because they yield comparable or better detection limits, and are simpler to use. They studied Cd, Ga, Pb,

Tl, Fe, and Sn. The intensity of fluorescence for the six elements was 2–3 times greater with a total consumption burner than with the chamber-type nebulizer, except for Fe where the signals were comparable. Turbulent flames, did, unfortunately, produce more intense thermal emission than the premixed flames, as illustrated in Table 5.

The vaporization of solvent in turbulent flames is quite incomplete, especially in the low regions of the flame. The high rise velocity for the sample droplets through the turbulent flame further decreases the degree of

TABLE 5

RELATIVE INTENSITY OF ATOMIC FLUORESCENCE AND OF
THERMAL EMISSION SIGNALS IN PREMIXED AND
TURBULENT FLAMES

	Premixed flame		Turbulent flame	
Element	Atomic fluorescence	Thermal emission	Atomic fluorescence	Thermal emission
Ga 4172 Å	0.32	0.00	1.00	0.57
Pb 4058 Å	0.28	0.17	1.00	0.22
Tl 5350 Å	0.29	0.02	1.00	0.29

vaporization. Premixed flames with chamber-type nebulizers yield increased atomic fluorescence signals with organic solvents, while with total consumption burners, on the contrary, the organic solvent introduced directly into the flame increases its background and noise and often adversely affects in the detection limits. Turbulent flames exhibit the phenomenon of scattering to a much greater extent than the premixed flames. This effect is due to the presence of larger size droplets of sample and the unvaporized solute particles in the turbulent flame. The effect of scattering is, however, compensated for by the increased sensitivity obtained with the turbulent flame.

Both turbulent and premixed flames are widely used, and, in spite of individual workers conviction of the superiority of one or the other configuration, both configurations yield excellent results.

3. *Burner Height*

The intensity of background radiation of all flames varies with different burner heights and decreases continuously as higher and higher regions of the flame are aligned with the entrance slit of the monochromator. Figure 6 illustrates the background variation as a function of burner height for air–hydrogen, for air–acetylene, and for air–propane premixed flames (20). It is clear that the background can be almost completely eliminated by taking measurements very high in the flame. However, in the high regions of the flame the temperature decreases rapidly, and some flames tend to fluctuate more, thus increasing noise. Thus the optimum burner height for a given flame and a given element must always be

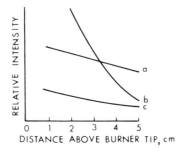

Fig. 6. Background intensity at 2750 Å in different flames: (a) air–acetylene (multiply ordinate by 16 to compare with curves b and c); (b) air–propane; (c) air–hydrogen (20).

determined on the basis of the best signal-to-noise ratio rather than just the intensity of fluorescence.

Easily atomized elements, such as Cd, Cu, Hg, Ag, and Zn, can best be determined above the visible part of the turbulent flame rather than within it. A position just above the tip of the inner cone has been found suitable for some elements (13, 14) in premixed propane flames. Elements such as Fe, Mg, and Ni, which are slightly more difficult to atomize, are often determined within the luminous portion of the flame and give best signal intensities in the region of maximum temperature. However, it must be remembered that, even for the more refractory elements whose emission is greatest from the region of highest temperature, this region may not be a practical choice for analysis if the background has strong features at the analytical wavelength. For example, Mg cannot be determined with best precision in the hottest part of the flame due to the intense OH background emission at 2852 Å. The optimum burner height in this case would be some experimentally determined compromise between the hottest part and the region completely above the flame.

Bratzel and Winefordner (41) optimized burner heights to most favorable signal-to-noise ratios for many elements and found that for turbulent hydrogen flames best fluorescence signals were obtained at 2–4 cm above the luminous part of the flame. Thus, using a Beckman total-consumption burner, Ag, Cu, and Hg were best measured at 10 cm above the burner tip, while Au, Bi, Co, Mg, Mn, Pb, Se, Te, Tl, and Zn gave best results at heights between 7 and 9.5 cm. In most cases, the region of observation was well above the region of maximum temperature, and for many elements this position did not correspond to the most intense fluorescence signal, but only to the most favorable signal-to-noise ratio.

4. Separated Flames

Separated flames are known in flame emission work and their use has been extended to atomic fluorescence (42, 43). Separation is accomplished by a silica tube or by a stream of nitrogen flowing parallel to the flame. A normal nitrous oxide–acetylene flame has the advantage of efficient atomization of refractory elements, but exhibits a high and noisy background. When such a flame is separated and the expanded interconal region of the flame is viewed, the intensity of background emission is greatly decreased, while its high temperature and reducing character are retained. West and co-workers (42) have determined atomic fluorescence of beryllium in this flame and reported a detection limit of 0.01 μg/ml (43). The air–acetylene flame separated by a parallel stream of nitrogen yielded good results for Zn and Cd, with detection limits of 0.0002 and 0.0005 μg/ml, respectively (44).

The decreased intensity of background emission in a separated flame presents a great advantage when an unmodulated source is employed. When the source is modulated, the tuned detector is not sensitive to direct emission from the flame anyway, so that a decrease in background intensity in this case serves only to decrease the noise. However, efficient atomization in the hot and reducing environment of the interconal zone increases the sensitivity of the analysis regardless of the nature of the source.

5. Nonflame Cells

Although a flame is almost exclusively the sample cell employed in atomic fluorescence, and all remaining sections of this chapter refer to a flame cell, two recently reported nonflame cells must be mentioned. In a heated graphite cell (45) samples are vaporized in an argon atmosphere. The method yields the absolute detection limits given in Table 6.

TABLE 6

Element	Detection limit, g
Zn	4×10^{-14}
Cd	2.5×10^{-13}
Sb	2×10^{-10}
Fe	3×10^{-9}
Tl	2×10^{-9}
Pb	3.5×10^{-11}
Mg	3.5×10^{-12}
Cu	4.5×10^{-10}

Another nonflame cell which has been described is a carbon filament (46). In this approach, sample solutions are placed on the filament and vaporized in an argon atmosphere. The reported detection limits are 3×10^{-11} g for Ag and 10^{-16} g for Mg.

C. OPTICS AND MONOCHROMATION

Collimating and converging lenses located between the source and the flame and between the flame and the monochromator do not present any real advantages. The same or better sensitivity is achieved simply by bringing the source as closely as possible to the flame (6, 13). When this is done, however, special care must be taken to prevent any light from the source from entering the monochromator by the use of baffles along the sides of the lamp. Also, if the distance between the lamp and the flame is of the order of a few centimeters, care must be taken to shield the lamp sufficiently from the heat of the flame in order to preserve its current stability. Bringing the flame as close as possible to the entrance slit of the monochromator also improves sensitivity (13).

The entrance and exit slits of the monochromator are adjusted to give the best signal-to-noise ratio for each element under the given flame conditions. This often results in keeping the slits rather wide and, if necessary to reduce sensitivity, in decreasing the amplifier gain, since high setting of amplification gain can generate added noise (13). When a flame of high background intensity is used, however, slit widths have to be reduced. Fluorescence is directly proportional to slit width, but the background varies as the second power of slit width (20). Thus, the signal-to-background ratio benefits from narrow slit widths.

The intensity of fluorescence is proportional to the total absorption which takes place in the analytical region Y of the flame (Fig. 2). Increasing the solid angles over which the excitation occurs and over which the monochromator views the flame would increase the total absorption, but will also increase the amount of stray light and noise reaching the detector. The total absorption can also be increased by passing the excitation beam through the flame several times. A concave mirror placed opposite the

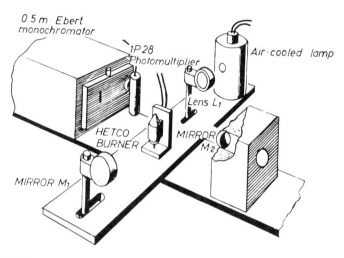

Fig. 7. Mirror arrangement around the flame. [Reprinted from (*33*) by courtesy of Elsevier.]

source reflects the exciting beam through the flame and increases fluorescence intensity. A concave mirror placed behind the flame, in the optical path of the monochromator, reflects additional fluorescence upon the entrance slit of the monochromator. The two mirrors together increase the detected signal intensity by about 168% (*33*). Figure 7 illustrates the mirror arrangement.

Any monochromator suitable for flame emission work is also suitable for atomic fluorescence as long as its capability extends well into the uv region. The function of the monochromator in atomic fluorescence work is the isolation of the desired fluorescence line from other fluorescence lines of the element of interest, from fluorescence lines of other elements which may also be excited by the given source and mainly, in an un-modulated system, from the thermal flame background at the surrounding wavelengths. However, many elements can be atomized efficiently in the relatively cool flames like air–hydrogen, which have a very low background emission intensity, especially at burner heights of several centimeters above the visible part of the flame. In such cases, it is clear that sophisticated monochromation systems are not only of no advantage but may be a serious handicap due to unavoidable loss of light intensity, and may seriously limit the sensitivity of the determination. Instead, a simple interference

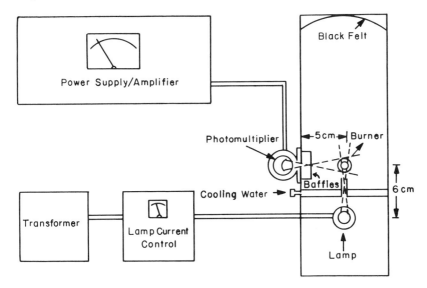

Fig. 8. Nondispersive instrumental arrangement for atomic fluorescence work. [Reprinted from (47) by courtesy of American Chemical Society.]

filter with a wide bandpass is the best choice. Good monochromation is necessary only in the very rare cases when actually spectral interference due to line overlap occurs. And, of course, with an unmodulated source, the more luminous the flame, the more necessity there is for good monochromation.

With low background flames, sensitive measurements of atomic fluorescence are possible without the use of any monochromator at all. Figure 8 illustrates schematically the nondispersive system which was used to determine Cd, Zn, and Hg (47). The reported detection limits, obtained with spectral discharge lamps as excitation sources, were 0.001 μg/ml for Cd, 0.0002 μg/ml for Zn, and 1.0 μg/ml for Hg. A nondispersive atomic fluorescence determination of iron has also been reported (48). Whether the fluorescence spectrum of the element consists mainly of one intense line, as is the case for Cd and Zn, or whether it consists of many lines, the total detected signal remains indicative of the sample concentration of the element of interest. Clearly, the nondispersive method is not useful in the presence of spectral interference. A nondispersive atomic fluorescence instrument would be rugged, cheap, and easy to operate but at the cost of a great loss in versatility.

IV. Working Curves and Detection Limits

Experimental working curves conform to the general shapes predicted by theory (Fig. 4). The linear portions of the curves often extend from the

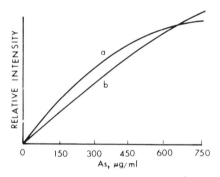

Fig. 9. Experimental calibration curves for As at 1972 Å with an electrodeless discharge tube as the source: (a) nitrogen (entrained air)–hydrogen flame and (b) air–acetylene flame. [Reprinted from (24) by courtesy of Pergamon Press.]

detection limits to concentrations 1000 or more times higher. This large range of linearity, together with good detection limits, are the most useful aspects of atomic fluorescence spectrometry. Several experimental calibration curves are illustrated in Figs. 9–11.

As would be expected, the extent of the linear working range depends markedly upon the composition of the flame. Table 7 illustrates the effect of three different turbulent flames upon the linearity of the working curve for magnesium (39).

Table 8 lists the best detection limits reported to date in atomic fluorescence. Since definitions of detection limits differ among laboratories, some of the listed values may vary by as much as a factor of 2 or 3 if

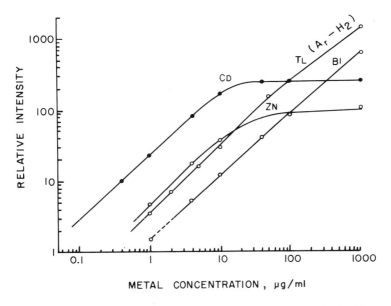

Fig. 10. Experimental calibration curves for several elements obtained with a continuous source (150-W xenon arc). Oxygen–hydrogen flame for Cd, Bi, and Zn and argon (entrained air)–hydrogen flame for Tl. [Reprinted from (*21*) by courtesy of American Chemical Society.]

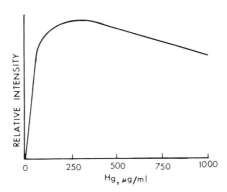

Fig. 11. Experimental calibration curve for mercury at 2537 Å with an electrodeless discharge tube as the source and air–hydrogen flame (*34*).

TABLE 7

EFFECT OF THE NATURE OF THE FLAME UPON THE
EXTENT OF THE LINEAR WORKING RANGE OF
MAGNESIUM

Flame	Linear range, $\mu g/ml$
Air–H_2 not premixed	0–1
Air–H_2 premixed	0–5
O_2–H_2 not premixed	0–20
O_2–H_2 premixed	0–5
N_2O–H_2 not premixed	0–2
N_2O–H_2 premixed	0–2

reduced to a consistent definition. All the data are for aqueous solutions
and most were obtained with electrodeless discharge tubes as sources, but
for different optimum flame conditions. These detection limits compare
quite favorably with those of atomic absorption and flame emission, and
may be expected to improve further with improvements in instrumentation,
especially the excitation sources.

V. Interferences

Interference effects observed in atomic fluorescence can be divided into
three groups: spectral interferences, physical interferences, and chemical
interferences (36).

A. SPECTRAL INTERFERENCES

Spectral interference occurs when another element emits fluorescence
radiation simultaneously with the test element and both wavelengths fall
within the bandpass of the monochromator. With most line sources, such
interferences are relatively rare and, usually, even rare enough to make
nondispersive atomic fluorescence possible.

Instances of excitation of a fluorescence line of an element by a spectrally overlapping line of another element include the excitation of the Cd
resonance line at 2288.02 Å by the As 2288.12-Å line (23), the excitation
of the 2061.70-Å Bi resonance line by the 2061.63-Å iodine line (23), and
the excitation of Fe, Tl, Cr, and Mg lines by Hg emission lines (33).

TABLE 8
ATOMIC FLUORESCENCE DETECTION LIMITS

Element	Fluorescence wavelength, Å	Detection limit, μg/ml	Reference
Ag	3281	0.0001	(25)
As	1937	0.13	(49)
Au	2676	0.2	(25)
Be	2349	0.01	(42)
Bi	3025	0.025	(23)
Ca	4227	0.02	(25)
Cd	2288	0.000001	(25)
Co	2407	0.005	(50)
Cr	3572	0.05	(49)
Cu	3247	0.001	(32)
Fe	2483	0.008	(49)
Ga	4172	0.3	(49)
Ge	2652	10	(25)
Hf	2866	3^a	(25)
Hg	2537	0.08	(34)
In	4105	0.1	(25)
Ir	2544	>1000	(28)
Mg	2852	0.001	(14)
Mn	2794	0.006	(25)
Mo	3798	2	(25)
Na	5892	100	(30)
Ni	2320	0.003	(31)
Pb	4058	0.02	(32)
Pd	3404	2	(28)
Pt	2659	50	(28)
Rh	3692	3	(28)
Ru	3728	200	(28)
Sb	2176	0.05	(23)
Sc	3907	10^a	(25)
Se	1961	0.15	(23)
Si	2516	0.55	(51)
Sn	3034	0.1	(52)
Sr	4607	0.03	(25)
Te	2143	0.05	(23)
Ti	3949	6^a	(25)
Tl	3776	0.008	(25)
U	3812	5^a	(25)
Zn	2139	0.00004	(25)
Zr	3520	4^a	(25)

[a] It is possible that these signals were a result of scattering of exciting light (8).

If analysis is based on a spectrally overlapping line in the presence of the interfering element, erroneous enhancement of the signal will be observed.

Spectral overlap does not always constitute an undesirable coincidence. As was pointed out before, the Hg lamp can be conveniently used to analyze for Fe, Tl, Cr, and Mg, and the iodine lamp has been employed for the determination of Bi. The obvious requirement in these applications is that the lamp element be absent from the analysis sample.

Thermal emission of the same lines as the fluorescence ones also falls in the category of spectral interference. For example, Table 9 lists the

TABLE 9

Flame	Relative fluorescence intensity	Relative thermal emission intensity
Air–hydrogen	1.57	0.04
Oxygen–hydrogen	1.43	0.16
Nitrous oxide–hydrogen	1.50	0.06

relative fluorescence and thermal emission signals obtained for Mg in different premixed flames optimized for fluorescence (*39*). Both thermal emission and flame background emission are eliminated if a modulated source and detector are used. However, when the flame background is very weak, and a cool hydrogen flame is used, modulation is not usually necessary because thermal emission by atoms in these flames is very weak below 3000 Å, where most fluorescence lines are found.

B. PHYSICAL INTERFERENCES

Physical interference characteristic of atomic fluorescence is due mainly to scattering of the exciting radiation by solvent droplets and unvaporized solute particles in the flame. The solids most likely to be present in unvaporized state to a significant extent are the compounds of the salt introduced into the flame or the refractory compounds involving the element and flame gas products like O, or OH, which form in the flame.

The turbulent flames obtained with a total consumption burner result in a more pronounced scattering of the exciting radiation than premixed flames, due to larger droplets and to more particles of unvaporized salts. Scattering is progressively more pronounced as the point of observation is lowered into the visible portion of the flame and approaches the primary reaction zone (Fig. 12). The amount of incident light scattering can be estimated by measuring the intensity of the signal resulting at the nearby line emitted by the source, assuming that the instrument response is the same at both wavelengths (40). The effect of scattering can be corrected for with continuum excitation sources by scanning the region of the fluorescence line and by means of zero suppression. The drawback of this

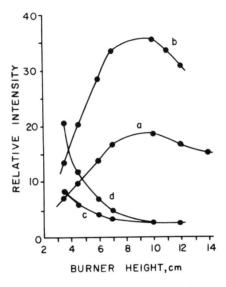

Fig. 12. Atomic fluorescence of cadmium and scatter in premixed and turbulent nitrous oxide–hydrogen flames obtained with an integral sprayer burner: (a) atomic fluorescence of cadmium in premixed flame; (b) atomic fluorescence of cadmium in turbulent flame; (c) scatter in premix flame; (d) scatter in turbulent flame. [Reprinted from (39) by courtesy of American Chemical Society.]

approach is that continuous sources have relatively low intensities in the near uv where many of the most useful fluorescence lines are located. While it can help to detect the presence of scattering qualitatively, it is incorrect to estimate the degree of scattering by aspirating a solution of another element which gives refractory compounds into the flame and by assuming that the amount of scattering will be the same. It will not be the same because the degree of scattering depends on the element tested and the anion among other parameters (40).

Scattering of the continuous radiation of a Xe arc by salt particles results in the detection of broad bands whose intensity is proportional to the concentration of these particles in the flame. The intensity of the scattering signal also depends, to a much smaller degree, on the nature of the element and on the nature of the compound in which it was introduced into the flame. Scattering curves are useless for identification but yield fair detection limits for individual elements alone. A few detection limits are listed in Table 10 (*21*). Due to their lack of specificity the scattering curves have little analytical usefulness.

TABLE 10

DETECTION LIMITS BASED UPON SCATTERING OF
CONTINUOUS RADIATION

Element	Wavelength of measurement, Å	Detection limit, μg/ml
Ca	4227	1.5
Ba	4000	7
Ga	4680	20
Ni	3370	13

Solvent droplet size and their contribution to scattering is minimized by using a spray-chamber aspirator-burner and by substituting an organic solvent for water.

C. CHEMICAL INTERFERENCES

The chemical interferences observed in atomic fluorescence are the same as those which affect both of the other flame methods. Any interactions which decrease the atomic population of the element of interest will decrease the intensity of atomic fluorescence, as will any interactions which involve the excited state of the atoms and inhibit their normal radiative decay.

Data are accumulating on the effects on atomic fluorescence of many elements by 1000- or 100-fold excesses of various cations and anions added to the sample. For Cd and Zn no interferences were found in air–propane although 41 cations and 18 anions were tested, including such common interferents as Al, Fe, Mo, phosphate, silicate, and oxalate (*20, 53*). The only element which is found to depress the Ag fluorescence signal is Al (*13*).

The Mg and Ca fluorescence signals from an (entrained air)–hydrogen flame are depressed by a great number of foreign ions (*36*). All common anions, except ClO_4^- and Cl^-, cause very significant depression. For example, a 100-fold excess of SiO_3^{2-} depressed both Ca and Mg fluorescence completely, while PO_4^{3-} depressed Ca completely and Mg by 77%. Al, Cr, Pb, Fe, and many other cations also cause great decreases in the Ca and Mg signals. Releasing agents, like EDTA, La, and Sr, employed in the other flame methods are useful in atomic fluorescence also. In the instances of Ca and Mg, making the sample solution 0.1% in $SrCl_2 \cdot 2H_2O$ eliminates the depressive effect of almost all foreign ions.

When large amounts of Sr, La, or EDTA are added to the sample for the purpose of eliminating chemical interferences, it must be remembered that doing so increases the probability of scattering of the exciting radiation by the high concentration of the foreign salts in the flame. Use of blanks corrects for the intensity of this scattering, but increased noise affects sensitivity (40).

The depressive effect on fluorescence of Mg by Al, Mo, and Ti, which form refractory oxides in the cooler flames, is less pronounced in air–acetylene and is not observed at all in the hot, reducing nitrous oxide–acetylene flame. Similarly, the interference of Fe is eliminated by turning to the acetylene-containing flames (14). Similar results were obtained when the influence of various inorganic acids and different cations upon the fluorescence signal of Fe, Co, and Ni were studied (31). In air–hydrogen flame the interferences were many and severe, all depressive, while in the air–acetylene flame they were almost completely eliminated.

The effects of 100- and 1000-fold excess of a large number of ions upon the Be fluorescence were tested in a separated nitrous oxide–acetylene flame. Interference was observed with 1000-fold excess of Al, Ca, and Mo, while a 100-fold excess of these ions gave no interference (42).

VI. Organic Solvents

It is commonly observed in the conventional flame methods, especially thermal emission, that substitution of organic solvents for water increases the intensity of the analytical signal. This beneficial effect is attributed to better nebulization and evaporation of the solvent. Effects of organic solvents in atomic fluorescence have not yet been studied in much detail. It has been shown, however, that the intensity of fluorescence is enhanced by organic solvents when premixed flames are used in conjunction with spray-chamber type burners. In the analysis for Ag, for example, the detection limit improved by a factor of 42 in an organic solvent (13).

When methyl isobutyl ketone (MIBK) was substituted for water, the detection limit for Zn improved by a factor of five (20), and that for Se improved from 0.2 to 0.08 $\mu g/ml$ (54). And the extraction of complexes of Fe, Co, and Ni into MIBK improved detection limits by factors of 100, 25, and 30, respectively (31). Figures 13 and 14 illustrate the effect of isopropanol upon Cd and Tl signals (40).

The effect of the organic solvents with turbulent flames produced with total consumption burners seems to be quite different (Figs. 13 and 14).

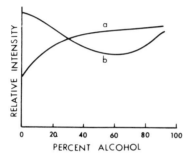

Fig. 13. Effect of isopropyl alcohol upon cadmium atomic fluorescence in premixed and turbulent air–hydrogen flames: (a) premixed and (b) turbulent (40).

Fig. 14. Effect of isopropyl alcohol upon atomic fluorescence of thallium in premixed and turbulent argon (entrained air)–hydrogen flames: (a) premixed and (b) turbulent (40).

With isopropyl alcohol, as illustrated, the signals for Cd and Tl go through a pronounced minimum. Similar effects were seen with ethyl alcohol.

The effects of methanol, ethanol, propanol, acetone, dioxane, acetic acid, and glycerol on atomic fluorescence of Fe, Co, and Ni were also investigated, and it was found that, although with a premixed air–hydrogen flame all organic solvents except glycerol caused enhancement of the fluorescence signal, in a turbulent flame the use of these solvents decreases the intensity of fluorescence (31).

VII. Evaluation of Atomic Fluorescence Spectrometry

Atomic fluorescence is an extremely specific and sensitive new method of analysis. It complements the conventional flame methods and it may equal them in importance, when the full potential of the new technique is realized.

Very few applications of atomic fluorescence spectrometry have been reported as yet. Jet engine lubricating oils have been analyzed for Ag, Cu, Fe, and Mg by aspirating the oils directly into the flame without any pretreatment (55). Nickel has been determined in petroleum fractions (56). Silicon content in low alloy steels was analyzed in a separated nitrous oxide–acetylene flame (57).

The most striking characteristic of atomic fluorescence is its great sensitivity. The detection limits in atomic fluorescence are usually at least an order of magnitude better than the "concentration for 1 % absorption" values listed in atomic absorption spectrometry. Although these two magnitudes are not directly intercomparable, it is clear that atomic fluorescence is at least as sensitive and often more so than atomic absorption when one compares the lowest concentrations commonly detectable by the two techniques. The important advantage in atomic fluorescence is that sensitivity can be drastically improved by improved source intensity.

Similar to atomic absorption, atomic fluorescence spectrometry is much less dependent upon flame temperature than is flame emission. It is most useful in the analysis of elements which emit their resonance lines in the uv where thermal emission methods are least sensitive. In addition, elements with weak absorption lines, like Hg 2537 Å, can be analyzed very sensitively by atomic fluorescence due to the elements' high quantum efficiency. For elements like the alkalies which emit strongly in the visible, flame emission is vastly superior. Also, elements which are especially difficult to atomize give relatively poor detection limits in atomic fluorescence. Compared to atomic absorption, atomic fluorescence has the disadvantage of being more susceptible to quenching by self-absorption and by collisions and reactions with various flame species. However, neither quenching, nor the light scattering interference effects have yet been exhaustively investigated.

Atomic fluorescence calibration curves have a conveniently large range of linearity which gives this technique a better dynamic range than that of atomic absorption. For most elements, the slope remains unchanged from the limit of detection to a concentration 1000 or 10,000 times greater than the limit of detection.

The instrumentation required for atomic fluorescence work is usually not much more complex than that in the conventional flame methods. Only very rudimentary monochromation is required in many applications. Electrodeless discharge tubes are less expensive than the hollow-cathode tubes, and are much easier to manufacture. Source stability is more important in atomic fluorescence work than in atomic absorption where a double-beam optical arrangement is often employed. Although continuous sources are much more applicable to atomic fluorescence than to atomic absorption, their use does not constitute a significant advantage because continuous sources have invariably given less sensitive results than the electrodeless discharge tubes.

Atomic fluorescence spectrometry possesses great sensitivity and versatility and will undoubtedly become a widely used method for trace analysis.

REFERENCES

1. A. C. G. Mitchell and M. W. Zemansky, *Resonance Radiation and Excited Atoms*, Cambridge Univ. Press, New York, 1934 (reprinted in 1961 by Macmillan, New York).
2. E. L. Nichols and H. L. Howes, *Phys. Rev.*, **23**, 472 (1924).
3. J. W. Robinson, *Anal. Chim. Acta*, **24**, 254 (1961).
4. C. T. J. Alkemade, in *Proceedings Xth Colloquium Spectroscopicum Internationale* (B. F. Scribner and M. Margoshes, eds), Spartan, Washington, 1963, pp. 143–170.
5. J. D. Winefordner and T. J. Vickers, *Anal. Chem.*, **36**, 161 (1964).
6. J. D. Winefordner and R. A. Staab, *Anal. Chem.*, **36**, 165 (1964).
7. J. D. Winefordner and R. A. Staab, *Anal. Chem.*, **36**, 1367 (1964).
8. J. D. Winefordner and T. J. Vickers, *Anal. Chem.*, **42**, 206 R (1970).
9. J. D. Winefordner, M. L. Parsons, J. M. Mansfield, and W. J. McCarthy, *Spectrochim. Acta*, **23B**, 37 (1967).
10. J. D. Winefordner, M. L. Parsons, J. M. Mansfield, and W. J. McCarthy, *Anal. Chem.*, **39**, 436 (1967).
11. S. J. Pearse, L. de Galan, and J. D. Winefordner, *Spectrochim. Acta*, **23B**, 793 (1968).
12. D. N. Armentrout, *Anal. Chem.*, **38**, 1235 (1966).
13. T. S. West and X. K. Williams, *Anal. Chem.*, **40**, 335 (1968).
14. T. S. West and X. K. Williams, *Anal. Chim. Acta*, **42**, 29 (1968).
15. J. I. Dinnin and A. W. Helz, *Anal. Chem.*, **39**, 1489 (1967).
16. G. Rossi and N. Omenetto, *Talanta*, **16**, 263 (1969).
17. J. M. Mansfield, J. D. Winefordner, and C. Veillon, *Anal. Chem.*, **37**, 1049 (1965).
18. G. I. Goodfellow, *Anal. Chim. Acta*, **36**, 132 (1966).
19. R. M. Dagnall, T. S. West, and P. Young, *Talanta*, **13**, 803 (1966).
20. R. M. Dagnall, K. C. Thompson, and T. S. West, *Anal. Chim. Acta*, **36**, 269 (1966).
21. C. Veillon, J. M. Mansfield, M. L. Parsons, and J. D. Winefordner, *Anal. Chem.*, **38**, 204 (1966).
22. D. W. Ellis and D. R. Demers, *Anal. Chem.*, **38**, 1943 (1966).
23. R. M. Dagnall and T. S. West, *Appl. Opt.*, **7**, 1287 (1968).
24. R. M. Dagnall, K. C. Thompson, and T. S. West, *Talanta*, **15**, 677 (1968).
25. K. E. Zacha, M. P. Bratzel, J. D. Winefordner, and J. M. Mansfield, *Anal. Chem.*, **40**, 1733 (1968).

26. J. M. Mansfield, M. P. Bratzel, H. O. Norgordon, D. O. Knapp, K. Zacha, and J. D. Winefordner, *Spectrochim. Acta*, **23B**, 389 (1968).
27. V. P. Pechorin and B. V. L'vov, *Zh. Prikl. Spektrosk.*, **7**, 764 (1968).
28. D. C. Manning and P. Heneage, *At. Abs. Newsletter*, **7**, 80 (1968).
29. D. W. Ellis and D. R. Demers, 153rd National Meeting, American Chemical Society, Miami, Florida, April 1967.
30. J. I. Dinnin, *Anal. Chem.*, **39**, 1491 (1967).
31. J. Matousek and V. Sychra, *Anal. Chem.*, **41**, 518 (1969).
32. D. C. Manning and P. Heneage, *At. Abs. Newsletter*, **6**, 126 (1967).
33. N. Omenetto and G. Rossi, *Anal. Chim. Acta*, **40**, 195 (1968).
34. R. F. Browner, R. M. Dagnall, and T. S. West, *Talanta*, **16**, 75 (1969).
35. B. W. Bailey, *Spectroscopy Letters*, **2**, 81 (1969).
36. D. R. Demers and D. W. Ellis, *Anal. Chem.*, **40**, 860 (1968).
37. R. Smith, C. M. Stafford, and J. D. Winefordner, *Anal. Chem.*, **41**, 946 (1969).
38. R. M. Dagnall, K. C. Thompson, and T. S. West, *Anal. Chim. Acta*, **41**, 551 (1968).
39. M. P. Bratzel, R. M. Dagnall, and J. D. Winefordner, *Anal. Chem.*, **41**, 1527 (1969).
40. M. P. Bratzel, R. M. Dagnall, and J. D. Winefordner, *Anal. Chem.*, **41**, 713 (1969).
41. M. P. Bratzel and J. D. Winefordner, *Anal. Letters*, **1**, 43 (1967).
42. D. N. Hingle, G. F. Kirkbright, and T. S. West, *Analyst*, **93**, 522 (1968).
43. G. F. Kirkbright and T. S. West, *Appl. Opt.*, **7**, 1305 (1968).
44. R. S. Hobbs, G. F. Kirkbright, M. Sargent, and T. S. West, *Talanta*, **15**, 997 (1968).
45. H. Massmann, *Spectrochim. Acta*, **23B**, 215 (1968).
46. T. S. West and X. K. Williams, *Anal. Chim. Acta*, **45**, 27 (1969).
47. T. J. Vickers and R. M. Vaught, *Anal. Chem.*, **41**, 1476 (1969).
48. P. L. Larkins, R. M. Lowe, J. V. Sullivan, and A. Walsh, *Spectrochim. Acta*, **24B**, 187 (1969).
49. R. M. Dagnall, M. R. G. Taylor, and T. S. West, *Spectroscopy Letters*, **1**, 397 (1968).
50. B. Fleet, K. V. Liberty, and T. S. West, *Anal. Chim. Acta*, **45**, 205 (1969).
51. R. M. Dagnall, G. F. Kirkbright, T. S. West, and R. Wood, *Anal. Chim. Acta*, **47**, 407 (1969).
52. R. F. Browner, R. M. Dagnall, and T. S. West, *Anal. Chim. Acta*, **46**, 207 (1969).
53. T. S. West, in *Trace Characterization* (W. W. Meinke and B. F. Scribner, eds), NBS Monograph 100, Washington, 1966, pp. 284–299.
54. M. S. Cresser and T. S. West, *Spectroscopy Letters*, **2**, 9 (1969).
55. R. Smith, C. M. Stafford, and J. D. Winefordner, *Can. Spectry.*, **14**, 2 (1969).
56. V. Sychra, J. Matousek, and S. Masek, *Chem. Listy*, **63**, 177 (1969).
57. G. F. Kirkbright, A. P. Rao, and T. S. West, *Anal. Letters*, **2**, 465 (1969).

9 Commercial Instruments

John A. Dean

DEPARTMENT OF CHEMISTRY
UNIVERSITY OF TENNESSEE
KNOXVILLE, TENNESSEE

I. Introduction

The steadily growing demands for chemical analyses of various materials ranging from the depths of the ocean to rocks from the moon has compelled the analyst to seek new instrumentation to make the analyses simple, sensitive, and precise. To fill this void created in our technological advancement, flame emission and atomic absorption spectrometry have surged forward to a new era. Although the evolution of atomic absorption is still proceeding at a very rapid rate, as indicated by the introduction of new components and sales of commercial instruments, atomic absorption has not eliminated flame emission but rather has complemented it. Similarly, atomic fluorescence has not nor will not replace absorption or emission. The state of the art of these methods has grown to where 10^{-14} g can be detected in a sample of only a few milligrams or a major constituent (5 to 100 %) can be determined with a relative standard deviation of 0.2 %.

Today the analyst has a wide variety of commercial instruments and modular components from which to choose. Light sources, burners and nebulizers, optics, and electronics are described in detail in preceding chapters. It is not the intent of this chapter to evaluate or describe all of the instruments that are commercially available. However, an effort has been made to summarize some of the basic instrumental components as given in Appendix 1.

Many atomic absorption instrument designs utilize a single-beam, ac system. The design trend in instrumentation is toward modular-compact units, simplicity in operation, increased sensitivity, and digital computer technology for acquisition of data (1-4). The modular-compact unit concept has arisen as a means of instrumentation for different competences and as a space saver in the laboratory. With solid state electronics and integrated circuits the operation of an emission or absorption instrument has been simplified with increased sensitivity. More of the instrumentation is being designed toward automation. Instruments are available for the analysis of Na and K which on command are able to standardize, dilute the sample, compute the results, print out the data, and shut down the operation of the instrument. It is conceivable that in the near future instruments can be built to be controlled by a computer so as to prepare the dilution, check for interferences, and report the data.

II. Commercial Spectrometers for Absorption, Emission, and Fluorescence

A. AZTEC

The Atomic Analyzer is designed to analyze metals by the three analytical techniques of atomic absorption, atomic emission, and atomic fluorescence. This permits the operator to choose the analytical method best suited to his particular problem, and to achieve maximum sensitivities and detection limits. The basic instrument incorporates both atomic absorption and atomic emission capability; switching from one to the other involves merely operating two switches on the front panel. Atomic fluorescence is available as an optional accessory, which includes electrodeless discharge lamps, a microwave generator, and a cavity. The

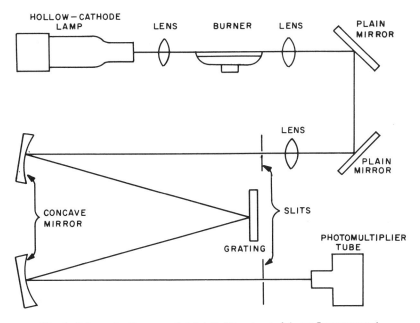

Fig. 1. Schematic diagram of AAA-3. (Courtesy of Aztec Instruments.)

optical rail pivots at the center line of the burner, and can be readily rotated a full 90° to the normal optical path for atomic fluorescence operation. All electronics are completely solid state and are contained on five printed circuit boards. They are readily accessible for easy servicing and replacement.

An outstanding feature of the instrument is the monochromator. In order to conserve valuable bench space, periscope optics are used (Fig. 1). The optical rail, on which source lamps, lens, and burner-nebulizer assembly are attached, is mounted on a rigid optical bench; the mono-chromator is bolted to the same bench from underneath; thus achieving distortion-free rigidity. The entire monochromator, periscope assembly and the photomultiplier are sealed and fitted with tubing connected to a nipple located on the back of the gas control unit. Thus, the entire optical section may be purged with dry gases. Purging with dry nitrogen will improve performance in the far uv for such analyses as As and Se. The monochromator employs a modified Czerny-Turner mount, 0.5-m focal length, and with $f/10$ optics. Slits are continuously variable from 5–100 μm. The grating used has 1200 grooves/mm, blazed at 3000 Å. Resolution is approximately 0.2 Å, and reciprocal linear dispersion is 16 Å/mm.

A lamp rack, holding up to three hollow-cathode lamps, rests on its optical rail saddle. The operating lamp is modulated at 400 Hz; the other two lamps are in warm-up position at either 5 or 10 mA. There is an extra connection so that three additional lamps may be mounted on a second lamp rack and allowed to warm up before using. The lamp rack will accommodate any hollow-cathode lamp of either the $1\frac{1}{2}$-in. or 2-in. diam varieties.

Another important feature of the basic instrument is a motor-driven scanning mechanism, controlled by switches on the front panel. Scanning speeds of 10, 20, 50, 100, 500, and 1000 Å/min are possible, and a 4000-Å/min speed is incorporated when the operator wishes to traverse from one wavelength range to another. This feature adds qualitative capabilities to the model AAA-3. With an accessory recorder, the operator can scan the entire spectrum from the far uv through the visible, and record emission features in unknown samples.

The atomic emission mode is achieved merely by means of a single switch on the front panel of the instrument. This switch activates a mechanical tuning fork whose tines operate at 400 Hz, in synchronous operation with the amplifier.

B. BAUSCH AND LOMB

The AC2-20 atomic absorption spectrometer features a high dispersion grating monochromator. The monochromator has a dialable slit selector permitting choice of 2-, 5-, or 20-Å resolution with a filter selector for removal of second-order spectral lines. The electronics utilizes a high gain, stable, low drift detection system which is specifically designed for a single-beam system. The hollow-cathode lamp emits a modulated light output which gives rise to an ac signal from the photodetector. The power supply also puts out another ac signal in phase with the lamp signal to trigger the synchronous demodulator in the spectrometer. The frequency of modulation is 133 Hz.

The optical system utilizes a unique arrangement of the hollow-cathode lamps. Three lamps are arranged normal to the main axis. A mirror-lens combination rotates from one detent position to another to pick up light from the desired lamp and direct it down the optical axis of the flame. The first lens images the bright cathode spot in the center of the flame. The second lens relays the cathode image to the entrance slit of the monochromator.

C. Beckman

This company offers a choice of models: 440, 444, and 448. The model 444 may be operated as a double- or single-beam instrument. In the double-beam mode the source is monitored, which permits immediate operation without a warm-up period. Some of the outstanding features of model 440 and 444 include a digital display of data, recorder output, computer compatibility, and single- or triple-pass optics (1).

All models can be equipped with automatic sample changers and data handling devices such as printers, recorders, or tape punch with teletype for computer calculation. A choice of burners is offered which include the laminar flow with heated spray chamber, the Autolam for handling solutions with high concentration of solids and the turbulent burner.

D. Bendix

The model A3000 has both flame emission and atomic absorption capabilities. This instrument is produced in England by Southern Analytical and distributed by the Bendix Corporation in the United States after some modification. See Section M for a description of the instrument.

E. Carl Zeiss

The model FMA is capable of flame emission or atomic absorption spectrometry. It is of a modular design arranged on an optical bench. Due to the separation into components, the instrument can easily be adapted to many different tasks or extended further. The monochromator can be equipped with a high-quality prism or grating. A double monochromator is available for special tasks.

The radiation emitted by the hollow-cathode lamp in the absorption module is modulated with a mechanical chopper and passes through a lens into the burner compartment. To effectively increase the path length of the burner, the radiation is reflected twice through the flame by means of mirrors. The radiation leaves the flame compartment through two lenses and thence passes into the monochromator. A stream of compressed air serves as a protective sheath for the flame as well as the means of nebulization, a unique feature of the burner. With the protective sheath the analyst can operate even when laboratory atmospheres are very unfavorable. An optional automatic cutoff offers protection against faulty operation by interrupting the supply of fuel when oxidant is absent.

For the emission mode the absorption unit is removed and replaced with the emission module. This unit uses a turbulent sprayer burner capable of using oxygen–hydrogen or oxygen–acetylene gases.

F. HEATH

The Heath Company has introduced a series of spectrophotometric instrumentation utilizing a concept of modular units that can be conveniently integrated with the Heath "700" monochromator into systems of instrumentation including an atomic absorption, emission, or fluorescence spectrometer as well as other spectrometric instrumentation. Conversion from one instrumental system into another is accomplished by means of a self-aligning instrument base which locks the associated modules into a single unitized instrument.

The central module is the Heath "700" monochromator with 0.35-m Czerny-Turner grating mount; the wavelength is read out on a 5-digit counter. The slits are bilateral, continuously adjustable between 5 and 2000 μm in width, and variable in height between 0.5 and 12 mm.

The instrumental design provides for convenient conversion from absorption to emission or fluorescence modes. Both hollow-cathode and vapor-discharge lamps are used as light sources. An arrangement of limiting apertures and/or lenses permits selection of variable size areas of the flame to be observed. Atomic absorption and atomic fluorescence systems can be operated in both modulated and unmodulated modes. In the modulated mode, electronic circuitry integrates and averages signals to provide for accurate subtraction of flame noise and photomultiplier dark current.

The large number of independent variables available to the user makes this instrumentation especially useful for academic and industrial research; also, the system is equally well suited for routine use in control laboratories by technicians through step-by-step instructions presented in the same lucid style familiar to users of other Heath equipment.

G. HILGER AND WATTS

This company manufactures the "atomspek" atomic absorption spectrophotometer shown in Fig. 2. This instrument covers the wavelength range from 1930 to 8530 Å. It has a built-in, 6-lamp turret with a multi-lamp power supply designed to keep up to four lamps ready for immediate use. Other components include the monochromator, detector, 3-stage amplifier, and meter readout. The first amplification stage is tuned at 400 Hz, the same frequency at which the hollow-cathode lamps are modulated. The final amplification stage is logarithmic in order to provide a linear absorbance readout—available on meter or external recorder. Wavelength and slit settings are displayed digitally.

H. INSTRUMENTATION LABORATORIES

This company has introduced a new concept into atomic absorption spectrometry. Model 353 is a dual channel/double beam system which provides two separate and distinct absorption/emission systems which utilize a common burner. This optical system supplies the capability for two-element analyses but also enables the use of internal standardization. The hollow-cathode light beam is split to allow one-half of the light to pass through the flame and the other one-half to by-pass the flame. The signal that by-passes the flame is used as a reference to correct for any drift from the hollow-cathode lamp. The second channel permits internal standardiza-tion which is a means of correcting for some types of chemical and physical interferences (2, 5). The burner system of these models uses a premixed counter flow jet of oxidant and fuel to increase the efficiency of nebuliza-tion about threefold.

I. JARRELL-ASH

Four models of atomic absorption spectrometers are currently available. They are the Dial-Atom (82-720), Atomsorb (82-270), Maximum Ver-satility (82-500), and the "800" (82-800). Each model offers distinct features specifically designed for the analyst.

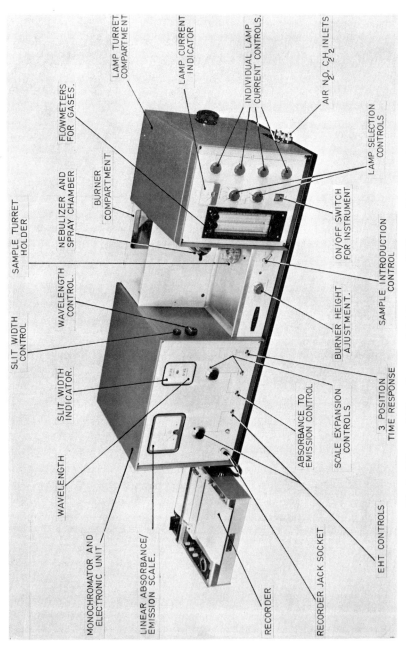

Fig. 2. Hilger and Watts "atomspek." (Courtesy of Hilger and Watts Ltd.)

The Dial-Atom II is the most compact and economical instrument produced by Jarrell-Ash. Its emphasis is on simplified operation. Nevertheless, it still has the features which are considered necessary for good analytical results. The unique performance features of the instrument include: (1) a hollow-cathode lamp holder which holds two lamps, each with independent current regulation, (2) wavelength selection by the chemical symbol as well as a nanometer scale, (3) conversion to the flame

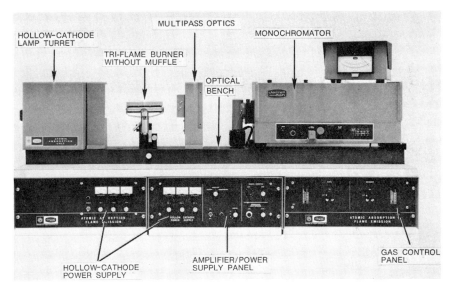

Fig. 3. Jarrell-Ash Maximum Versatility. (Courtesy of Jarrell-Ash, Division of Fisher Scientific Co.)

emission mode by merely flipping a switch, (4) interchangeable readout meter scales in transmittance, absorbance, percent absorption or concentration, and (5) interchangeable fixed slits.

The Atomsorb is specifically designed to meet all routine analytical requirements. The optical bench design and modular solid state electronics offer considerable flexibility. Other features of the Atomsorb include: (1) a turret holding six hollow-cathode lamps with individual power controls and meters for three of the lamps, (2) two back-to-back gratings blazed at 2500 Å and 6000 Å, respectively, each 64 × 64 mm with 1180 grooves/mm, and (3) a safety gas control for switching the oxidant from air to nitrous oxide while the flame is burning.

The Maximum Versatility (Fig. 3) is most suitable for research and other

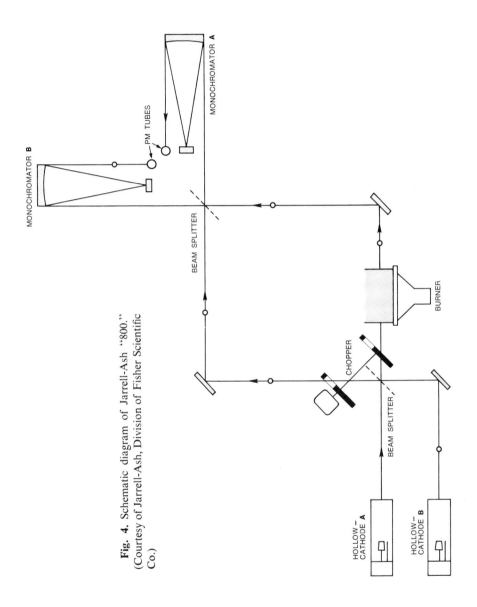

Fig. 4. Schematic diagram of Jarrell-Ash "800." (Courtesy of Jarrell-Ash, Division of Fisher Scientific Co.)

sophisticated applications. Every aspect of the instrument is built to allow maximum flexibility. The optical components are aligned on an open optical bench. The solid state electronics are of a modular concept adding to the versatility. Significant features of the instrument include: (1) a six lamp turret with a separate power control and meter to each hollow-cathode lamp, (2) a laminar flow and integral sprayer-burner system capable of using acetylene or hydrogen as fuel with air, nitrous oxide, or argon (entrained air) as the oxidant, (3) a 0.5-m monochromator with interchangeable gratings and several types of slits, (4) a multipass optical system that passes the beam of light five times through the flame, (5) a concentration mode in addition to the absorption and emission, (6) a damping control which can be inserted during aspiration without affecting the readout stability, (7) an optional electrical wavelength scanning drive with 8 speeds and accurate to within ± 1 Å, and (8) a gas control panel with two separate fuel connections and two separate oxidant connections.

The "800" consists of a dual-double beam optical system (Fig. 4). With each monochromator independent of the other, the analyst has available a number of unique choices: (1) two independent channels with each analyzing a separate element simultaneously; (2) one channel operating in the emission mode while the other operates in the atomic absorption mode; (3) one channel operating in the conventional absorption mode while the other channel determines the degree of background absorption at a nearby nonabsorbing line; (4) one channel operating in the conventional absorption mode while the other channel determines the degree of background absorption within a certain bandpass using a continuum light source; (5) a two-line method where each channel is tuned to two separate absorbing wavelengths of the same element; (6) where applicable, using one channel to measure the signal from an added internal standard while the other channel measures the unknown signal.

The analyst has also an option regarding the signal readout. Each channel can be read sequentially or simultaneously or as a ratio of the signals from each channel. The readout system includes an electronic digital display that enables the displayed signal to be integrated and averaged over a period of 1, 3, or 10 sec. The signal can be displayed as transmittance, absorbance, or concentration. If any slope curvature exists on the calibration curve, correction can be made at the point where the curvature begins. A feedback system to the photomultiplier power supply continually adjusts for any fluctuations in the light source giving an improvement in the signal-to-noise ratio and permitting greater scale expansion to be employed.

J. JOBIN-YVON

Jobin-Yvon, headquartered in France, manufactures two emission and absorption flame spectrophotometers: DELTA and VARAF II. DELTA is a single-beam instrument with a Czerny-Turner grating mount and slits adjustable both in height and width. The optical system is modulated mechanically at 125 Hz. An automatic wavelength drive has 10 speeds from 1 to 1000 Å/min. The lamp supply provides a regulated constant current for three hollow-cathode lamps simultaneously.

VARAF II, a flame photometer, which may be transformed easily to an atomic absorption spectrometer, has a grating monochromator covering the spectral range of 2000 to 8000 Å with spectral bandwith adjustable from 0.5 to 50 Å. The turbulent burner can be replaced with a laminar flow burner.

K. NORELCO-UNICAM

The Unicam SP90 has been designed to exploit the atomic absorption technique but still retain a capability for flame emission. A turret contains three hollow-cathode lamps which are operated independently, each at its correct operating current, thus avoiding lamp warm-up delays. By rotating the turret any chosen lamp is positioned without any further adjustment. The instrument incorporates a totally enclosed burner system which protects the flame from drafts and the operator from glare. An optional accessory is a multislot burner for use with high solids; up to 10% total solids may be handled without causing the head to clog. The monochromator has a 30° silica prism in a Littrow mount which covers the wavelength range from 2000 to 8500 Å. The slit width may be varied continuously between 0.01 and 2 mm by a cam and cam-follower mechanism. A system of shutters between the lamp and the flame, and between the flame and the monochromator entrance slit, allows rapid selection of either emission or absorption modes. In the emission mode a vibrating reed modulator, located behind the monochromator entrance slit, is activated for modulation of the light from the flame (6).

L. PERKIN-ELMER

The most widely used atomic absorption spectrometer is the Perkin-Elmer Model 303, a double-beam instrument, described by Kahn and

Slavin (7) (Fig. 5). The spectral range of the monochromator, from 1900 to 8700 Å, is covered with two gratings, positioned back to back in a 0.4-m Czerny-Turner mount. A rotating sector mirror alternately passes the beam from the source through the flame and then through a reference path by-passing the flame. The two beams are recombined by a mirror which transmits 50% of the light and reflects 50%. The ratio of the sample beam (through the flame) to the reference beam (by-passing the flame) is measured by the electronic system of the instrument. The readout is

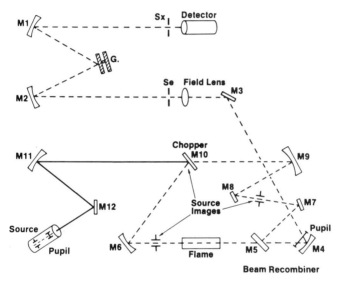

Fig. 5. Schematic diagram of the double-beam optical system of Perkin-Elmer model 303. (Courtesy of Perkin-Elmer Corp.)

determined by nulling the ratio signal with a potentiometer circuit and reading the per cent absorption on a digital counter.

Koirtyohann (8) showed that effects due to molecular absorption in the flame or to light scattered by refractory particles in the flame can be compensated by comparing the ratio of the apparent absorption using a hollow-cathode source to the absorption using a source of continuous radiation. Molecular absorption and light scattering will absorb an equal fraction of the light from both sources, so that the ratio of the two apparent absorptions will yield a signal proportional to the absorption of the atomic vapor alone. This has been adapted for the model 303. The background corrector has proven very useful in the determination of low metal concentrations in a complex matrix, particularly for elements whose resonance line lies in the far ultraviolet, such as Zn, Pb, As, Se, Te, and Cd.

In a less expensive single-beam spectrometer, model 290 (Fig. 6), the hollow-cathode lamp is driven by a voltage-regulated square-wave power supply. The burner and the hollow-cathode lamps are identical to those used in other models.

To take advantage of the three competences (absorption, emission, and fluorescence), models 403 and 305 have been introduced. The purpose of

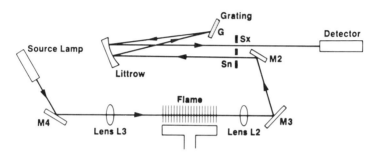

Fig. 6. Schematic diagram of Littrow mount in Perkin-Elmer model 290. (Courtesy of Perkin-Elmer Corp.)

the 403 is to take advantage of modern digital computer technology and procedures of data acquisition and reporting, Fig. 7. The optical system is very similar to Fig. 5 except that a deuterium discharge lamp is built directly into the instrument for use with the background correction system. In addition, a mechanical light interrupter is mounted inside the instrument between the flame and the monochromator. Actuation of the "chopper" coupled with a variable-speed wavelength drive-motor permits the direct recording of flame emission spectra. In the model 403, the circuit is entirely digital from the preamplifiers to the digital display and teletypewriter record. Integrated circuits are used throughout the digital readout. Each tenth second a complete reading is taken that is proportional to the concentration. Routinely, ten such readings are automatically averaged and displayed. However, if greater precision is required, 100 readings can be averaged in about 10 sec. Nonlinear working curves are accommodated with a circuit that can be adjusted very rapidly without requiring iterative steps. The zero signal is automatically adjusted by pressing an "automatic zero" button, while aspirating a sample blank.

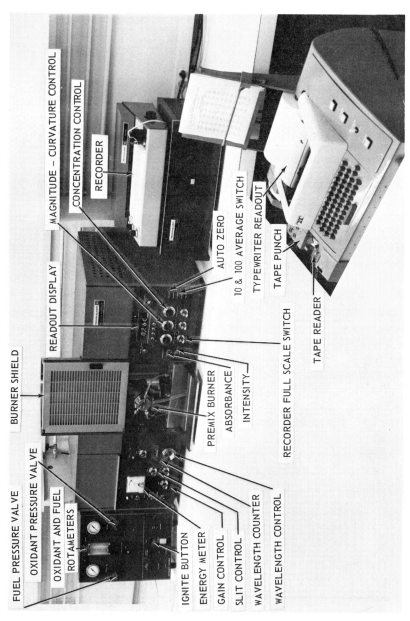

FUEL PRESSURE VALVE
OXIDANT PRESSURE VALVE
OXIDANT AND FUEL ROTAMETERS
BURNER SHIELD
MAGNITUDE – CURVATURE CONTROL
CONCENTRATION CONTROL
RECORDER
READOUT DISPLAY
AUTO ZERO
10 & 100 AVERAGE SWITCH
TYPEWRITER READOUT
TAPE PUNCH
PREMIX BURNER
ABSORBANCE
INTENSITY
RECORDER FULL SCALE SWITCH
TAPE READER
IGNITE BUTTON
ENERGY METER
GAIN CONTROL
SLIT CONTROL
WAVELENGTH COUNTER
WAVELENGTH CONTROL

Fig. 7. Perkin-Elmer model 403. (Courtesy of Perkin-Elmer Corp.)

The display can be recorded on a teletypewriter. This permits a great deal of versatility. The teletypewriter can be controlled either by the output of the model 403, from the keyboard itself, or by a punched paper tape in the tape reader of the typewriter. Thus, the operator can decide how he wishes to prepare the final report of the data and make a simple control tape to provide this format. If he wishes, he may type in the sample identification number as each sample is completed, or he may let the typewriter simply identify the samples sequentially.

M. SOUTHERN ANALYTICAL

Southern Analytical manufactures two instruments. Model A1740 is the only commercially available flame emission photometer which uses automatic background compensation. The spectrum is sampled by an exit slit which has a small mirror on either side. As the line of interest passes through the exit slit to a photodetector, the two sections of the spectrum on each side of the emission line are reflected onto a second photodetector. The mean value of the light intensity on either side of the spectral line is subtracted from the light intensity of the spectral line. The instrument can be used as a direct reading flame photometer or as a scanning flame photometer with electronic integration for periods ranging from 2 to 20 sec.

The atomic absorption spectrophotometer, model A3000, is a single-beam instrument. Hollow-cathode lamps are held in a four-position turret. When the lamp is moved into position on the optical axis, it is switched to a stabilized power supply while the three remaining lamps are held in a warm-up condition. Burners for air–acetylene, air–propane, or nitrous oxide–acetylene are available. The monochromator is a 0.25-m Czerny-Turner mount with a variable bandpass of 1.5 to 60 Å.

The electronic circuitry incorporates a logarithmic amplifier so that a linear readout in absorbance is obtained. Facilities are available for the connection of a chart recorder and digital voltmeter. The instrument can be used for flame emission and also for atomic fluorescence with hollow-cathode lamps or modulated microwave electrodeless discharge tubes as light sources.

The samples are introduced to the nebulizer by a rotatable turret, holding eight sample cups. The burner is designed to eliminate the danger of flashback and to allow flame ignition and turn-off without the necessity of burning pure acetylene or mixtures of air and nitrous oxide. All flames are ignited electrically from a front panel, and gas flows are controlled by needle valves and flowmeters. A double-walled chimney isolates the flame from the rest of the instrument. The rate of sample uptake is about 3 ml/min, which is less than usual with this type of aspirator.

N. SPECTRAMETRICS

This company has introduced a new dimension in spectrochemical analysis with an echelle grating-prism dispersion unit and a high-temperature plasma flame (9). A modified Czerny-Turner optical system is used to

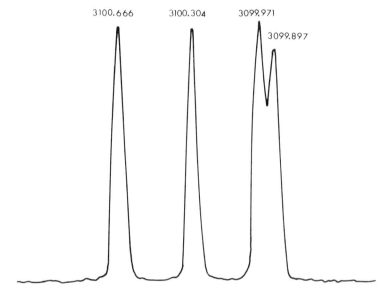

Fig. 8. Densitometer tracing of iron "triplet" at 3100 Å. Conditions: dispersion, 0.6 Å/mm in the 85th order; equivalent slit width, 100 μm; resolution, 60 mÅ; instrument resolving power, 500,000 (10-μm slit). (Courtesy of Spectrametrics Inc.)

provide minimum aberrations over the field of view required by an echelle grating and cross dispersion-prism combination (see Chap. 5, Section I.E). An example of the resolution of this instrument is given in Fig. 8.

Two instrument configurations are presently being offered. These are the model 101 selectable wavelength instrument and the model 201 simultaneous multiple wavelength instrument. These instruments are capable of being used either in the absorption or emission mode depending upon the type of sample excitation.

The standard type of excitation for atomic analysis is the SpectraJet (Fig. 9), a unique argon high-temperature plasma flame. Sample consumption is only 0.05 to 0.2 ml/min, enabling small amounts of sample to be handled. The SpectraJet requires flow rates of 5 to 10 ft³/hr of argon and approximately 300 W of electrical power. Optical background noise from the argon plasma is virtually eliminated by bending the plasma out of the field of view of the spectrometer. The atomic vapor of the sample does not bend and is observed above the argon plasma. In the emission

Fig. 9. SpectraJet. (Courtesy of Spectrametrics Inc.)

mode the light emitted by the excited atoms is optically chopped and then focused on the entrance slit of the spectrometer. In the atomic absorption mode, light from a continuum source is optically chopped and focused in the SpectraJet. Resolving power is sufficiently high that it provides selection of the desired wavelength without the use of a hollow-cathode lamp.

O. VARIAN TECHTRON

This company manufactures several models of absorption and emission equipment. Model AA-5, as shown in Fig. 10, is of a modular design with capabilities for absorption, fluorescence, and emission. Its great flexibility makes it ideal for the researcher or analyst with special problems. This instrument features automatic baseline correction which provides instantaneous adjustment of the baseline when drift occurs and when aspirating a blank solution which absorbs. The meter readout is either as $\%T$ or absorbance and has up to $10\times$ scale expansion in the absorbance mode. Outputs for a digital readout and recorder are available from the amplifier. The digital readout provides values in $\%T$, absorbance or concentration and offers scale expansion factors from $0.01\times$ to $20\times$. Signal averaging is continuously variable from 1 to 10 sec.

The instrument has a 0.5-m Ebert monochromator with high speed motorized scanning. The slits are continuously adjustable from 0 to 300 μm. The interior of the burner is made entirely of corrosion resistant materials, and five different burner heads are available to adapt for any common gas mixture.

The AA-5 may be readily adapted to the technique of atomic fluorescence by the addition of a second optical rail. High intensity lamps may be powered directly from the power supply. Additional sources such as electrodeless discharge lamps or continuums may be easily mounted on the second rail.

Model AA-120 is a compact modular instrument with about the same sensitivity as the AA-5. All components are mounted in plug-in boxes for ease of service and updating. Some of the features of the instrument are: a tilt-out meter, built-in burner ignition, a four-lamp turret, continuously variable slit height and width, built-in flame emission capability, and front panel burner controls.

Another member of the Techtron line is the AA-100. It is easy to operate and is capable of being operated on ac or a 12-V battery. Its

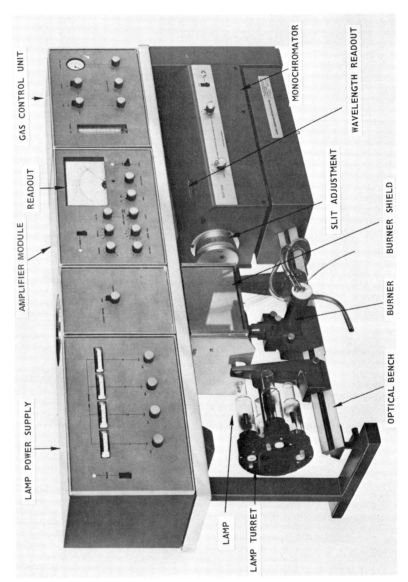

Fig. 10. Techtron model AA-5. (Courtesy of Varian Techtron Ltd.)

portability makes it ideal for remote locations or mobile field station use.

The AR-200 series employs a resonance detector instead of a mono-chromator to isolate the resonance lines (*10*).

III. Filter Flame Photometers

Instruments in this category are characterized by simplicity in operation and compactness in design (Fig. 11). Although able to determine up to six

Fig. 11. Clinical flame photometer. (Courtesy of Evans Electroselenium Ltd.)

or eight elements, these instruments are primarily used to determine sodium and potassium. Interference filters isolate the analytical line of interest. (See Chapter 5, Sections I.A and IV.B). With the trend toward automation many companies are redesigning their instruments to keep pace with modern technology. Instruments designed for automation include Electro-Synthese model PHF 62/A, Evans Electroselenium (EEL) model 170, and Yallen model 100. Samples are prepared and placed on turntables or prepared by automatic samplers which permits continuous operation. Standards may be tagged to automatically initiate restandard-ization of the flame photometer (EEL-170). Data is recorded, digitally displayed, and is capable of direct tie-in with line printers, tape punch computers, or with remote station operation. The last sample in a run can return the instrument to stand-by position.

Instruments that use twin channels with lithium as an internal standard include Advanced Instruments model 11D, EEL-170, Process Instruments model H-2, and Yallen model 100. A fixed quantity of lithium is added to the test solutions and standards, and emission intensity ratios are determined. The emission characteristics of lithium are very similar to those of sodium and potassium, which permits its use as an internal standard (11).

APPENDIX 1

Company	Model	Mode	Optics	Type of modulation, Hz	Nebu-lizer burner
1. Advanced Instruments Newton Highlands, Mass.	11D	FES	db	—	pm
2. Aztec Instruments Norwalk, Conn.	AAA-3	AAS FES AFS	sb	Pulsed-400	pm
3. Baird-Atomic Cambridge, Mass	KY-4	FES	db	—	pm
4. Bausch and Lomb Rochester, N.Y.	AC2-20	AAS	sb	Pulsed-133	pm
5. Beckman Instruments Fullerton, Calif.	440	AAS FES	sb	Mechanical	pm
	444	AAS FES	sb, db	Mechanical	pm
	448	AAS FES	sb	Mechanical	pm
6. Bendix Cincinnati, Ohio	A3000 (see Southern Analytical)				
7. Carl Zeiss Oberkochen, W. Germany	FMA	AAS FES	sb	Mechanical	pm, tb
8. Coleman Instruments Div. of Perkin-Elmer Maywood, Ill.	21 139- (Hitachi)	FES AAS FES	sb sb	— —	tb tb, pm
9. Electro-Synthese Arcueil, France	PHF 62/A	FES	sb	—	pm

Instruments which use single channels are the Coleman model 21, EEL model 100, and Electro-Synthese model PHF 62/A.

The early models of filter flame photometers used the integral sprayer burner. In recent years the trend has been toward premixed laminar-flow burners. The smaller bore solution capillary used in the sprayer burners clogs easily, and these burners were audibly noisy (Chap. 3). The Coleman model 21 uses a specially designed sprayer burner which produces a stable flame over a wide range of gas pressures. The fuel is diffused through a conical screen which contributes to an improvement in flame stability.

COMMERCIAL INSTRUMENTS

Dispersing unit, grooves/mm	Type of mount	Dispersion, Å/mm	Slits	Type of scan	Wavelength range, Å	Scale expansion in absorption	Readout system
1. Filter, Na, K (Li internal std)	—	—	—	—	—	—	Meter
2. Grating, 1200	0.5 m C-T	16	Vari	Elec	1850–9000	20×	Meter, recorder
3. Filter, Na, K (Li internal std)	—	—	—	—	—	—	Digital
4. Grating, 1200	0.5 m C-T	15	Fixed	Manu	1900–8500	10×	Meter, recorder
5. Grating, 1200	Littrow	20	Adj	Elec	1900–8520	10×	Digital, recorder
Grating, 1200	Littrow	20	Adj	Elec	1900–8520	10×	Digital, recorder
Grating, 1200	Littrow	20	Adj	Elec	1900–8520		Meter
6. (See Southern Analytical)							
7. Prism or grating	Littrow	—	Vari	Manu	2000–9000	4.5×	Meter, recorder
8. Filter, Li, Na, K, Mg, Ca	—	—	—	—	—	—	Meter
Grating, 1440	0.4 m Littrow	20	Vari	—	1950–8000	10×	Meter, recorder
9. Filter, Li, Na, K, Ca, Ba	—	—	—	—	—		Meter

Company	Model	Mode	Optics	Type of modulation, Hz	Nebulizer burner
10. Evans Electroselenium	EEL 100	FES	sb	—	pm
London, England	EEL 150	FES	db	Pulsed-50	pm
Div. of Corning Glass,	EEL 140	AAS) FES	sb	—	pm
Corning, N.Y.					
	EEL 170	FES	db		pm
	EEL 240	AAS) FES	sb	Pulsed-350	pm
11. Heath	703	AAS) FES AFS	sb	Mechanical	tb, pm
Benton Harbor, Mich.					
12. Hilger and Watts	Atom-spek	AAS	sb	Pulsed-400	pm
London, England					
13. Instrumentation	253	AAS FES	sb-db	Pulsed-A-500 B-1000	pm
Laboratories					
Lexington, Mass.	353	AAS) FES	Dual db	Pulsed-A-500 B-1000	pm
14. Jarrell-Ash	Dial-Atom	AAS) FES	sb	Pulsed-120	pm
Div. of Fisher					
Scientific	Atomsorb.	AAS	sb	Pulsed-120	pm
Waltham, Mass.	Max. Versat.	AAS) FES AFS	sb	Pulsed-120	pm
	"800"	AAS FES AFS	Dual db	Pulsed-120	pm
15. Jobin-Yvon	DELTA	AAS) FES	sb	Pulsed-125	tb, pm
Arcueil, France	VARAF II	FES (AAS)	sb	Pulsed-125	tb, pm
16. National Instruments	NIL	FES	db	—	pm
Rockville, Md.	101	AAS	sb	Pulsed-60	tb
17. Norelco-Unicam	SP90	AAS) FES	sb	Pulsed	pm
Philips Electronic Inst.					
Mount Vernon, N.Y.					
18. Perkin-Elmer	290B	AAS	sb	Mechanical	pm
Norwalk, Conn.	303	AAS	db	Mechanical	pm
	403	AAS) FES	db	Mechanical	pm
	305	AAS) FES	db	Mechanical	pm

(Continued)

Dispersing unit, grooves/mm	Type of mount	Dispersion, Å/mm	Slits	Type of scan	Wavelength range, Å	Scale expansion in absorption	Readout system
10. Filter, Li, Na, K	—	—	—	—	—	—	Meter
Filter, Na, K	—	—	—	—	—	—	Meter
Grating, 576	0.25m C-T	66	Vari	Manu	2000–10000	10 ×	Meter, recorder
Filter, Na, K (Li internal std)	—	—	—	—	—	—	Digital
Grating, 576	0.25m C-T	66	Adj	Manu	1800–8900	10 ×	Meter, digital, recorder
11. Grating, 1180	0.35m C-T	20	Vari	Elec	1950–7000		Meter, recorder
12. Silica prism	—	17* 45†	Vari	Manu	1930–8530	10 ×	Meter, recorder,
13. Grating, 1200	0.33m Eb	25	Adj	Elec	1900–10000	100 ×	Digital, recorder
Grating, 1200	0.33m Eb	25	Adj	Elec	1900–10000	100 ×	Digital, recorder
14. Grating, 1180	0.25m Eb	33	Fixed	Manu	1900–8600	20 ×	Meter, recorder
Grating, 1180	0.25m Eb	33	Fixed	Manu	1900–8600	20 ×	Meter, recorder
Grating, 1180	0.5m Eb	16	Adj	Elec	1900–8600	40 ×	Meter, recorder, digital
Grating, A-1180 B-2360	0.4m C-T	21 10	Adj	Elec	1850–10000	40 ×	Digital, recorder
15. Grating, 1200	0.6m C-T	12	Vari	Elec	1850–8500	5 ×	Meter, recorder
Grating	—	—	Vari	—	2000–8000	—	Meter, recorder, digital
16. Filter, Na, K (Li internal std)	—	—	—	—	—	—	Digital
Grating	Littrow	—	Vari	Manu	1850–9000	10 ×	Meter, recorder, digital
17. Silica prism	Littrow	3+ 32++	Vari	Manu	1900–8520	10 ×	Meter, recorder
18. Grating, 1800	0.27m Littrow	16	Adj	Manu	1900–8700	4.5 ×	Meter, recorder
Dual grating 2880 and 1440	0.4m C-T	6.5 13	Adj	Manu	1900–8700	25 ×	Digital, recorder
Dual grating 2880 and 1440	0.4m C-T	6.5 13	Adj	Elec	1900–8700	100 ×	Digital, recorder
Dual grating 2880 and 1440	0.4m C-T	6.5 13	Adj	Elec	1900–8700	100 ×	Digital, recorder

Company	Model	Mode	Optics	Type of modulation, Hz	Nebulizer burner
19. Process and Instruments, Corp. Brooklyn, N.Y.	H-2	FES	db	—	pm
20. Southern Analytical Camberley, England	A1740	FES	sb	—	tb, pm
	A3000	AAS FES	sb	Pulsed-400	pm
21. Spectrametrics Burlington, Mass.	101	AAS FES	sb	Mechanical	pt
	201	AAS FES	sb	Mechanical	pt
22. Varian Techtron Melbourne, Australia Division of Varian Associates, Palo Alto, Calif.	AA-5	AAS FES AFS	sb	Pulsed-285	pm
	AA-100	AAS	sb	Pulsed-285	pm
	AA-120	AAS	sb	Pulsed-285	pm
	AR-200	AAS	rd	Pulsed-285	pm
23. Yallen Brockton, Mass.	100	FES	db	—	tb

AAS = atomic absorption spectrometry
AFS = atomic fluorescence spectrometry
FES = flame emission spectrometry
db = double beam
sb = single beam

rd = resonance detector
pm = premix-laminar flow
tb = turbulent flow (total consumption)
C-T = Czerny-Turner
Eb = Ebert

REFERENCES

1. J. Ramirez-Muñoz, *Am. Laboratory*, August 1969, 25.
2. F. J. Feldman, *Am. Laboratory*, August 1969, 41.
3. H. L. Kahn, *Am. Laboratory*, August 1969, 52.
4. T. J. Poulos, *Am. Laboratory*, November 1969, 61.
5. F. J. Feldman, J. A. Blasi, and S. B. Smith, Jr., *Anal. Chem.*, **41**, 1095 (1969).
6. W. J. Price, *Spectrovision, Unicam*, **15**, 2 (1969).
7. H. L. Kahn and W. Slavin, *Appl. Opt.*, **2**, 931 (1963).
8. S. R. Koirtyohann, *Anal. Chem.*, **37**, 601 (1965).
9. W. E. Elliot, *Am. Laboratory*, March 1970, 67.
10. J. V. Sullivan and A. Walsh, *Appl. Opt.*, **7**, 1271 (1968).
11. J. A. Dean, *Flame Photometry*, McGraw-Hill, New York, 1960.

Dispersing unit, grooves/mm	Type of mount	Dispersion, Å/mm	Slits	Type of scan	Wavelength range, Å	Scale expansion in absorption	Readout system
19. Filter, Na, K (Li internal std)	—	—	—	—	—	—	Meter
20. Grating, 1180	0.25m C-T	32	Fixed	Elec	—	—	Meter, recorder
Grating, 700	0.25m C-T	60	Vari	Elec	1850–10000	10×	Meter, recorder
21. ec grating 73	0.75m C-T	2.5	Fixed	Elec	2000–20000	—	Meter
ec grating 73	0.75m C-T	2.5	Fixed	Elec	2000–20000	—	Digital
22. Grating, 638	0.5m Eb	33	Vari	Elec	1860–10000	100×	Meter, recorder, digital
Grating, 638	0.25m Eb	66	Fixed	Manu	1850–9500	10×	Meter, recorder, digital
Grating, 1274	0.25m Eb	33	Vari	Manu	1850–10000	10×	Meter, recorder, digital
—	—	—	—	—	—	—	Meter, recorder
23. Filter, Na, K (Li internal std)	—	—	—	—	—	—	Digital

Vari = variable   ec = echelle grating
Adj = adjustable   * = at 2000 Å
Elec = electric, motor-driven   † = at 5000 Å
Manu = manual   + = at 2000 Å
pt = plasma torch   ++ = at 4000 Å

10 Sample Preparation and Separation Methods*

D. W. Ellis

DEPARTMENT OF CHEMISTRY
UNIVERSITY OF NEW HAMPSHIRE
DURHAM, NEW HAMPSHIRE

I. Introduction

The kinds of samples suitable for analysis by flame emission, atomic absorption, or atomic fluorescence spectrometry range widely. The analyte may be present in only trace quantities, or it may be a major constituent.

* The work upon which part of this chapter is based was supported in part by funds provided by the United States Department of the Interior, Office of Water Resources Research, as authorized under the Water Resources Research Act of 1964.

263

The type and quantity of sample being analyzed usually affects the selection of a sampling method and the sample's preparation. Trace analysis will frequently require special sampling and preparation techniques. This chapter will be concerned with (a) factors affecting accuracy and precision, (b) the process of sampling; (c) procedures for obtaining the sample in solution, and (d) methods of separation or preconcentration.

Several authors in a number of different books and articles have addressed themselves to these subjects. The reader is referred to them for additional information and references (1–6), and useful references for different applications have been compiled (7). Reference (3) is particularly helpful as it contains many references from non-English journals.

II. Factors Affecting Accuracy and Precision

Both the precision and accuracy involved with sampling, sample preparation, and determination steps must come under scrutiny. Methods whereby the variability of an analysis can be broken down and assigned to different steps in the overall procedure have been described (8–12). The analyst must remember, however, that the overall precision implies accuracy only if the procedure has no systematic bias. In order to determine if this is true, "perfect" standards of the same physical state and composition as the original material must be available, and blanks must be carefully chosen. Standards with these qualities can sometimes be prepared from synthetic materials, for example, with metal alloys. In some cases (for example with some biological and geological samples) standards must be obtained from representative material. This involves careful selection and pretreatment of the material and analysis by many cooperating laboratories using methods which can be shown to be free of determinate errors. This approach has been followed recently for leaf material using kale (13) and for silicate rocks (14). The availability of many standard samples from the National Bureau of Standards (U.S.) has been of immense value in this regard. If such standards are not available, the analyst should evaluate every individual step of the analysis with respect to both precision and accuracy using radioactive tracers. The increased confidence in the results more than offsets the additional work required. The determinations of cobalt in blood (15) and cesium in plant and animal tissue (16) are good examples of how radioactive tracers have been used for this purpose.

The need for pure reagents and solvents to avoid contamination and the resulting potentially serious loss in accuracy is obvious, but attaining this goal can be extremely difficult. Water, the most commonly used solvent, can best be purified using ion-exchange resins (4) or distillation in quartz apparatus; contaminant ions in acids or other solvents are usually best removed with an appropriate chelating agent. Mizuike (4) provides a list of standard methods for the purification of commonly used acids and bases. Freeman and Kuehner (17) have discussed the preparation of several common reagents in ultrapure form. Organic solvents are usually freer of inorganic contaminants than acids. Other reagents, primarily fluxes, are frequently impure and in some instances are difficult to purify. Electrochemical techniques, zone refining, and recrystallization are all used to prepare pure standards and reagents. Robertson (18) has provided specific data on contamination levels in common reagents, solvents, etc., using neutron activation analysis and multidimensional gamma-ray spectroscopy.

When working at the trace level and in dilute solutions, all points of contact between a sample and any other material must be carefully scrutinized. Losses occur primarily by adsorption and lead to low levels of accuracy and some loss in precision. Contamination, which results primarily in poor accuracy, may occur by the desorption of a previously adsorbed species, by the decomposition of apparatus, or from reagents. Minczewski (3) has presented data on the effects of impurities in the air on selected techniques encountered in chemical analysis.

Several investigations have been concerned with a type of container which minimizes the effects of contamination and loss. Mitchell (19) has studied contamination from various types of containers; Minczewski (3) gives the following list in order of preference: Teflon and fluorinated plastics > polyethylene > quartz > platinum > glass. Most workers seem in agreement as to the undesirability of using glass as a storage vessel.

Eichholz and co-workers (20) in a more limited study with several radioactive isotopes (Ca, Sr, Y, Ce, Ba-La, Zr, I, and Ru-Rh) found that borosilicate glass was preferable to polypropylene for most of the elements studied. They also found that beaker coatings provided only partial protection against adsorption losses. Dyck (21), working with silver ion and "Desicote" as a glassware coating, found that plastics are preferable to glass only over a period of a few days and that plastics are generally more difficult to clean. West and co-workers (22) also worked with silver to evaluate surface losses and the effects of ligands as a means of keeping ions in solution.

Two other factors have a bearing on contamination and loss and hence on accuracy and precision; one is the ratio of solution volume to surface area and the second is the amount of time that the solution and container are in contact. The need to minimize the surface area with which the sample comes in contact is obvious. In addition, since equilibria reactions are involved, it is important to minimize the time of contact. One obvious conclusion is that dilute solutions should not be stored over long periods of time and that this is particularly true for standards. It is necessary for standards to be stored as relatively concentrated solutions, over short periods of time, and that standards of lower concentration must be prepared immediately before use from high purity solvents and must be discarded after perhaps one day's use. Since trace impurities are frequently found in diluents in varying quantities; blank determinations with just diluent should also be run as a matter of practice. Freezing has also been proposed and is being used at the National Bureau of Standards as a way of storing ultrapure reagents (17).

III. Sampling

The particular sampling approach to be followed should be dictated by the type of sample, the concentration of the element to be analyzed, and the specific purpose for the analysis. The sampling procedure or nonhomogeneity of the sample will sometimes introduce errors much greater than those associated with the chemical preparation or the determination of the sample. For example, Calder (8) has shown that in the analysis of potassium in herbage the variance due to sampling was ten times greater than the variances for the determination (flame photometric) and for the chemical preparation of the sample combined; for the manganese analysis, the variance due to sampling was 20 times larger. Data such as his must alert the analyst to be wary of assuming that he has a homogeneous sample or that his sampling procedure is satisfactory. This is not to exclude those cases where stratified sampling is required; nevertheless, within any stratum, the sample to be analyzed should be homogeneous. Naturally, sampling difficulties depend considerably upon the type of material being analyzed, biological samples being one of the more difficult. Difficulties are greatest when performing trace analyses but can occur also at the macro level.

IV. Obtaining the Sample in Solution Form

Depending on the sample matrix, a wide variety of dissolution steps are commonly used. Naturally, one method of sample dissolution will not work well for all matrices, and it is necessary to choose the best approach for any particular sample. Slavin (7) has reviewed the utilization of atomic absorption for a broad range of applications including biochemical, agricultural, metallurgical, mining and geochemical, industrial, and miscellaneous ones. In all cases he has provided an indication of the sample material and analytes; the appropriate references include information on sample dissolution. Each of the following methods are commonly used and represent reasonable approaches for many samples.

A. PRETREATMENT OF THE SAMPLE

Washing, drying, cleaning, grinding, and storage of a sample are all possible sources of loss or contamination. Though pretreatment is frequently necessary, one should be careful to evaluate its effect; the use of tracers or neutron activation analysis is extremely useful for this purpose. Steyn (23) and Chapman (24) have described representative difficulties as regards foliar samples.

B. ASHING

One of the standard methods for removing unwanted organic material has been dry ashing (5, 25); Gorsuch (26) has evaluated several different ashing procedures using radioactive tracers. In dry ashing, the sample is usually heated for several hours at approximately 500°C in the presence of air. Several difficulties with this technique are well known: (a) the loss of certain metallic constituents due to volatilization depending on the medium (1); (b) impurities introduced with the air (5, 15) or from the container; (c) drying and pre-ashing (15), which simply add to the time required; (d) with some samples, dissolving the ash is difficult. It should be added, however, that dry ashing is adaptable to a rather routine operation and hence is not particularly demanding of the analyst's time. Two commercial companies, Tracerlab and International Plasma Corporation, have introduced low temperature ashing devices based on the work of Gleit and Holland (27, 28). The sample is decomposed by the action of an electronically excited gas, usually oxygen, which has been excited by a radio-frequency source. The manufacturers claim that this approach is superior to both wet and dry ashing in that volatile elements are not lost and that ionic material and morphology are maintained. In at least one instance, however, incomplete recoveries have been obtained for arsenic, selenium, and mercury, and the extent of recovery in one case was dependent upon the matrix of the sample (29). The device has met with some use, but it is still too soon to be sure of its breadth of applicability and freedom from loss or contamination.

Wet ashing or digestion is frequently recommended in preference to dry ashing particularly for trace element analysis and for acceptable recoveries (25, 26). In this approach, the sample is usually heated with a mixture of HNO_3 and $HClO_4$ though other combinations of acids with H_2SO_4 are used for certain samples. The use of organic acids has been suggested as a means of improving the procedure (30, 31). With wet ashing, most losses are eliminated though now one is faced with obtaining pure, metal-free reagents and with minimizing the leaching of unwanted trace contaminants from container walls. In addition, wet ashing requires greater amounts of technician time, and extreme care must be exercised in the use of hot perchloric acid because of its tendency to explode under a variety of circumstances.

Both wet and dry ashing procedures have been shown to be unsatisfactory for many of the more volatile elements; Mizuike (4) lists fourteen metals which may be lost in dry ashing procedures and twelve metals which likewise may be lost due to volatilization with wet ashing. If one is interested in analyzing for one of these elements, then the ashing procedure must be carefully checked. If losses are observed, the method must be modified by lowering the temperature or by addition of specific reagents to minimize volatilization or by having a closed system, such as a Parr bomb. Both ashing methods suffer from an additional disadvantage in that extremely stable complexes are frequently formed in the ashing; this can lead to serious difficulties if, subsequently, extraction is anticipated to perform a preconcentration or separation.

The use of various oxygen bomb or flask techniques for the analysis of metallic species in organic matrices is well known (32). Various modifications of these procedures have recently been described: Fujiwara and

Narasaki (*33*) developed an ashing procedure carried out in a specially designed bomb to minimize losses due to volatilization and contamination due to reagents; Burroughs and co-workers (*34*) used a modified oxygen flask method in which a fluorocarbon was added to facilitate analysis of an organosilicon compound; Belisle and co-workers (*35*) designed a flow-through oxygen flask method which permitted handling larger samples than is possible with the standard Schöinger method.

C. DISSOLUTION

Dissolving most metals and alloys is not too difficult though finding the proper set of conditions is occasionally time consuming. Hydrochloric, sulfuric, and nitric acids are most frequently used; hydrogen peroxide and hydrofluoric acid are sometimes added for specific samples. Perchloric acid is also used in certain cases. Fusion with lithium borate has been described (*36, 37*). For many materials [silicates are a well-known example (*38*)], fusion is the common method of dissolution in order to bring the sample to a state where it can be dissolved. Fusion methods in general are more time consuming and frequently may involve special apparatus.

These methods when used for trace analysis have been shown to be inaccurate in many cases. It is almost impossible to obtain acids and fluxes which are completely free of trace inorganic ions, though with special care they can be purified satisfactorily.

Recently, Bernas (*39*) has described a method for the decomposition and dissolution of silicates which should have broad applicability for silicate analysis as well as for many other types of organic and inorganic samples. In his method, Bernas proposes using a specially designed Teflon decomposition vessel which eliminates dimensional instability and contact of the sample or reagents with any material other than Teflon. The sample is decomposed in 30–40 min at a temperature of 110°C using a small quantity of aqua regia as a wetting agent and then hydrofluoric acid. Boric acid is added to the reacted mixture, including precipitates, and then water is added; a clear solution was found to result. If this approach proves as convenient and inexpensive as this single paper suggests, it could point the way toward a much easier and more straightforward handling of many difficult samples. For trace analysis, the problem of obtaining pure reagents still remains, although several other difficulties have been eliminated.

D. Extraction of Ionic Material from Soils and Plant Material

Direct extraction of ionic species from soils and to a much lesser extent from plant material (plant materials are more frequently ashed) has been studied extensively for many years. Viets (40) separated extractable species into three categories: (a) those which are water soluble, (b) those which are exchangeable with weak exchangers, and (c) those which are held still more strongly and which are extractable using strong chelating or complexing agents. For species of the first type, water obviously is used for the extraction. Many different reagents have been used for the easily exchangeable cations, though ammonium acetate is by far the most frequently recommended. Strong chelating agents and organic acids are generally used in attempts to extract the most strongly held cations. Bedrosian (41) has reviewed many literature references for the extraction of five specific heavy metal cations. A broad-range extractant frequently recommended is 2.5% acetic acid and 1 N ammonium acetate.

V. Methods of Separation and Preconcentration

Separation or preconcentration or both are frequently required as part of flame spectrometric methods. Separation methods, as distinct from preconcentration steps, are normally utilized to separate the elements of interest from the matrix or from other interfering species. Preconcentration steps are performed when the sensitivity of the method is insufficient without the added step.

Due to the specific nature of flame spectrometric methods, it is possible in most instances to perform a general nonspecific separation and preconcentration step for one or several groups of metals without attempting to separate one element from all others in the sample and to concentrate it simultaneously. This general assumption, however, should not be taken as being absolute. In specific instances, it may be necessary to separate two elements of similar properties or, alternatively, chemical reagents may be added to overcome potential interferences.

In this section, three common methods of separation and preconcentration will be discussed: extraction, ion-exchange, and precipitation. In addition, several other techniques which are less used or are of more recent interest will be mentioned briefly.

A. EXTRACTION

Liquid-liquid extraction methods have been used extensively in conjunction with flame spectrometric techniques (42–45). These methods have numerous advantages; they are simple, relatively free of contamination, rapid, inexpensive, and have broad applicability. Trace and other minor constituents usually are extracted from the matrix material; but, occasionally, the matrix element is extracted leaving the minor metallic species in the original sample solution (46). Minczewski (3) provides tables indicating various separation methods for matrix and trace elements using extraction. Within certain limits, the specific extraction procedure selected is not important since the specificity of the total method is supplied by the spectroscopic measurement and not by the separation procedure. It is generally an advantage if the metal of interest resides finally in an organic solution, since the use of certain organic solvents has been shown to generally enhance the sensitivity of the flame methods (45, 47, 48). Even though the extraction of metal ions from organic solutions into water is possible, it has no advantage in almost all cases and so will not be discussed. The necessity of carrying a blank throughout the procedure exists here, just as with most of the other techniques of sample preparation.

Theoretical considerations pertinent to liquid-liquid extraction and to the extraction of metal chelates in particular have been discussed by a number of authors (49–55). The extraction of a metal ion from an aqueous phase into an organic phase depends upon the metal ion forming an uncharged species. This has been accomplished in a variety of ways: as a covalent compound, as a chelate, or as an association complex or ion pair. Of these, the most commonly used is the chelate approach. Chelate extraction depends on the distribution ratio for the metal ion in the aqueous phase and the chelate in the organic phase. The overall distribution then depends on four equilibria: (1) the distribution of the molecular chelating agent between the organic and aqueous phases; (2) the ionization of the chelating agent in the aqueous phase, which is usually pH dependent; (3) the formation of the metal chelate; (4) the distribution of the metal chelate between the two phases. Typical equilibria for the case of dithizone (HDz) are:

$$HDz_{(aq)} \rightleftharpoons HDz_{(org)}$$

$$HDz \rightleftharpoons H^+ + Dz^-$$

$$M^{n+} + nDz^- \rightleftharpoons M(Dz)_n$$

$$M(Dz)_{n(aq)} \rightleftharpoons M(Dz)_{n(org)}$$

Under ideal conditions, the final distribution ratio for the metal in the aqueous and organic phases ($D = [M(Dz)_n]_{org}/[M^{n+}]_{aq}$) is a function of the chelating agent and its concentration, the organic solvent, the pH of the aqueous solution, and the rate of chelate formation, which is usually fast. Complicating factors (56–58) including association polymerization and surface phenomena sometimes make the system deviate from simple theory as when the distribution ratio varies with the metal ion concentration. Under conditions where D is constant, even at low concentrations, single or multiple extractions provide a means of obtaining an extremely good separation and preconcentration with a minimum of contamination and inconvenience.

With most chelating agents, several or many elements are extracted simultaneously. For many flame-photometric analyses, this attribute is used to advantage. In some instances, however, it is desirable to extract only one specific element of interest. This can frequently be accomplished by the proper use of masking agents, such as cyanide, citrate, tartrate, and EDTA, or by controlling other parameters such as pH (3, 52, 54, 55a).

Lastly, Blaedel and Haupert (59) have described a pseudo "extraction" technique using ion-exchange membranes. The separation capability is due primarily to the high selectivity attainable with modern ion-exchange membranes. Preconcentration is enhanced by the presence of a high ionic strength or by anionic exchangers in the extractant solution. With a relatively simple apparatus, 100-fold enrichments were obtained with losses of less than 1%. The time required was of the order of 2 hr, a slight disadvantage, and the authors are careful to point out other limitations. Nevertheless, this method should be worthy of further study.

Frequently used chelating agents include acetylacetone, 8-quinolinol (oxine, 8-hydroxyquinoline), dithizone (diphenylthiocarbazone), cupferron (ammonium salt of m-nitrosophenylhydroxylamine), 2-thenoyltrifluoroacetone (TTA), and several dithiocarbamate derivatives, particularly diethyldithiocarbamate and ammonium pyrrolidine dithiocarbamate (APDC). Freiser (60) lists many of the common extractable metal ions and extraction systems; Zolotov and Kuz'min (45) list chelating reagents with the test element, material analyzed, solvent, pH, and the sensitivity. Several new types of chelating agents have been proposed including hydroxytriazines (61) and thiothenoyltrifluoroacetone (62).

Ammonium pyrrolidine dithiocarbamate is receiving increasing attention as a broadly applicable chelating agent for use with many metallic elements (1, 44, 63–77). The ability of APDC to work well in acid solution and without different buffers is a significant advantage when APDC is compared with other chelating agents. Dithizone and oxine have traditionally been used as broad range chelating agents and both of these are still used extensively (*78–83*) as is cupferron, though it requires more different pH conditions for different elements (*84, 85*).

Among other types of extraction systems, several deserve mention. Extraction of metal chlorides and nitrates has been known and used for many years (*46, 86, 87*). Alkylphosphoric acids have been used for the extraction of In, Ga, Tl, Sb, and Bi (*88*), and the extraction of various metal iodides has been described (*89–91*). Ion-association systems and liquid ion exchangers are being used slightly for analytical separations (*92–97*). Ruthenium and osmium can be extracted as their tetroxides into carbon tetrachloride (*98*) or into molten tin (*99*). A β-diketone has been used for the extraction of lanthanides (*100*) and heteropoly acids have been used in an indirect method for phosphorus and silicon (*101*).

For purposes of analyzing specific materials or classes of compounds, the review by Zolotov and Kuz'min (*45*) provides an excellent starting point and many good references. Slavin (*7*) also gives a number of excellent references and Joyner et al. (*70*) list solvent-extraction systems of proven or possible use for concentrating trace elements in sea water. Numerous articles in the *Atomic Absorption Newsletter* have dealt with extraction as part of flame spectrometric methods.

Irving and Williams (*51*) have reviewed many of the practical aspects of liquid-liquid extraction, including a discussion of the effect of changing the volumes of the different phases. Also discussed are the various types of apparatus which are commonly used. For most extractions, a single-stage or "batch" extraction is performed in which the aqueous sample to be extracted is mixed in a separatory funnel with an immiscible organic phase plus chelating agent, if appropriate. The mixture is shaken, either manually or with a mechanical shaker, until equilibrium is obtained. In some cases, the shaking must be very gentle or a continuous extractor must be used to avoid foaming or the formation of an emulsion. Alternatively, centrifugation may be used to speed up the separation of layers. Repeated batch extractions may sometimes be required to insure removal of the species of interest. When a considerable time might be required for the equilibration or when the analyst wishes to work with as small a quantity of extractant as possible to enhance preconcentration, the use of a continuous extractor has distinct advantages. Although "back-extractions" or "backwashing" are needed with some methods, these are not usually required with flame methods.

B. Ion-Exchange

Ion-exchange methods have likely found their greatest application in the chromatographic separation of individual elements within groups of elements difficult to separate, such as the actinides and lanthanides. With analysis by flame spectrometry, such a high degree of selective separation is not usually required; rather the analyst is more likely interested in group separations, in the elimination of an interfering species, or in the removal of an excess of matrix ion. For these purposes, ion-exchange methods are well suited.

Most ion-exchange methods use solid ion exchangers; some liquid ion exchangers have also been used, but these behave more similarly to extraction and hence were mentioned briefly in that section. Solid ion exchangers in the form of resins are most frequently used. Most resins are made from high molecular-weight organic polymers, are insoluble in almost all solvents, but are permeable to ions. Incorporated into the resin are functional groups which provide the ion-exchange capability; cation-exchange resins (the cation is the mobile ion) are usually made of sulfonated styrene-divinylbenzene copolymers while anion-exchange resins (the anion is the mobile ion) have the same styrene-divinylbenzene copolymer but contain quaternary ammonium groups. Various other types of ion exchangers are also available; these include certain specially treated celluloses, some inorganic ion exchangers, and resins containing chelating groups (55a).

Operation of an ion-exchanger depends upon built-in fixed charges which are neutralized by mobile charges. The type of exchanger is categorized by the charge of the mobile ion. For most work with metallic ions, strong and weak acid exchangers are used; in some cases, however, complexes of metallic ions are separated using anion exchangers. Examples include halide complexes as well as anion complexes with sulfate, oxalate, and EDTA. A typical strong acid ion exchanger can exist in acid form, represented as $R^- \! - \! H^+$ or it may exist in some other form, such as $R^- \! - \! Na^+$. The separation of other ions from a solution in contact with the resin is dependent upon the equilibrium established between the ions and the resin, as symbolized by the following expression:

$$x(R^- - H^+) + M^{x+} + xNO_3^- \rightleftharpoons R_x^{x-} - M^{x+} + xH^+ + xNO_3^-$$
$$\text{(solid)} \qquad \text{(solution)} \qquad \text{(solid)} \qquad \text{(solution)}$$

Quantitatively, the ability of an ion exchange resin to attract and hold a specific ion is expressed by the distribution coefficient K_d, which is equal to the amount of the ion per gram of dry resin divided by the amount of the ion per milliliter of solution: $K_d = [M]_r/[M]$.

In the use of ion-exchange resins for separation and preconcentration, several other factors must also be taken into consideration. One is the capacity of the resin, usually expressed as milliequivalents of ion-exchangeable sites per gram of resin, another is the selectivity of the resin, a third is the rate of exchange, and finally, the reversibility of the process (*55a, 102–105*).

Both column and batch operation are used. Batch operation is simple, requires no special apparatus and is frequently more rapid than column operation. Column operation is preferred for almost all applications where one element must be separated from one or more other elements and tends to have much in common with distillation. Whichever approach is used, the sample solution is brought into contact with the ion exchanger, sufficient time for equilibration is permitted, and the solute(s) of interest is removed from the resin. For most flame spectrometric methods, the ion exchanger is used to separate and concentrate more than a single element, and hence a batch approach is usually satisfactory. In this case, the sample solution is brought into contact with the resin, the solution is stirred and filtered, and the metallic ions are then eluted with a strong acid. Difficulties with contamination from the resin or the eluting solvent are not uncommon and require caution. Losses due to adsorption or incomplete reversibility on the exchanger are also possible. An alternative approach to elution and to thereby avoid these difficulties is to ash the resin containing the ion of interest and to then analyze the ash.

Practical considerations in the use of ion exchangers include swelling of the exchanger, particle size, adsorption, the eluting solvent, which may be aqueous, nonaqueous, or a mixture of the two, and the use of complexing agents. In some applications there are additional practical considerations. The chemical stability of the exchanger or the physical properties of the exchanger, e.g., in radioactive solutions, may be extremely important.

Many applications of ion exchangers for separation and preconcentration of minor constituents are reported in the literature. Anion exchangers are frequently useful for determining trace impurities in metals or alloys where halide anion complexes are utilized. Govindaraju (*106*) has discussed an ion-exchange dissolution method for silicates which is applicable to analyses for boron, silicon, and trace metallic impurities. Strelow and co-workers have described a number of selective separations using cations exchangers for the separation of alkali metals from the alkaline earths and other elements (*107*) and have used anion-exchange chromatography to separate cadmium from silver and other elements (*108*). Strelow (*109*) has separated barium from strontium and other elements by using a mixed hydrochloric acid–organic solvent eluent and cation-exchange chromatography. Andersen and Hume (*110*) used a cation-exchange resin and organic complexing agents in the eluent to separate and concentrate strontium, barium, and other species in sea water. Perdue et al. (*111*) utilized organic complexing agents in the eluent to good effect with both cation- and anion-exchange resins. Chelating ion-exchange resins have also found limited use (*112, 113*).

Several other approaches utilizing ion-exchange mechanisms have also been shown to be workable. Tera, Ruch, and Morrison (*114, 115*) developed a method which they called precipitation ion exchange. In this approach, both the matrix and metallic impurities are placed on a column of cation exchanger and the matrix is precipitated by the addition of a suitable reagent, such as hydrochloric acid dissolved in an organic solvent. Traces can then be nonselectively eluted, and the authors report that as many as 40 trace elements from a variety of matrices may be preconcentrated in greater than 90% yield with only minor traces of the matrix element. Variations of the organic-hydrochloric acid eluent is one means to affect which matrix elements are precipitated, and this determines which elements will show a delayed breakthrough from the column. A technique using specially manufactured ion-exchange resin-loaded filter paper disks has been described (*116*), and its application to the analysis of trace impurities in molybdenum (*117*) and in tungsten and tungsten trioxide (*118*) has been demonstrated. For specific separations and preconcentrations, the ion exchangers which incorporate chelation sites have proved useful. For example, Beichler (*119*) has described a 20-fold concentration of six elements in industrial waste water using Dowex A-1; atomic absorption spectrometry was utilized in the analytical measurement.

One other method should be mentioned briefly for it has several interesting aspects, particularly for selective separations. Korkisch (*120*) has described a method which he calls combined ion exchange–solvent extraction. In this approach, a conventional ion exchanger coupled with a mixed solvent system containing a compound capable of acting as a liquid ion exchanger is used. This approach has been shown to provide distinctly better separations than were achieved with either ion exchange or extraction alone. Korkisch and Orlandini (*121*) have described one application involving the separation of thorium from rare earths and other elements.

C. PRECIPITATION

Various methods involving precipitation have been used for separation and preconcentration. Direct precipitation, coprecipitation or carrier precipitation, and cocrystallization are familar techniques. Though difficulties are associated with each, one of these methods can generally be used for a specific separation or preconcentration.

Direct precipitation has usually been used for separating the matrix element from other species in solution; specific recommendations for this purpose have been provided by Minczewski (3). For purposes of determining a major constituent by one of the flame methods, this step is usually unnecessary. However, this procedure is more often used as a means for removing an interfering matrix element prior to determining minor constituents. In these instances, coprecipitation may produce significant losses. In order to minimize coprecipitation, it is usually recommended that a method involving precipitation from homogeneous solution be used.

Direct precipitation is seldom used for purposes of precipitating only the minor constituents in solution. Difficulties with solubility, and the ultimate limit determined by the solubility product, supersaturation and the formation of colloidal suspensions all mitigate against this approach. A better approach for separating and concentrating trace and micro amounts is coprecipitation. This phenomenon involves physical entrainment, mixed crystal formation, and adsorption. (These are generally lumped together under the term coprecipitation for purposes of this discussion.)

Coprecipitation methods are based on the principle that many minor constituents of interest will coprecipitate in conjunction with a more plentiful precipitate. This approach has been shown to work well in the microgram range, but the coprecipitation may be incomplete in the milligram range (122). In some cases, it is sufficient to simply precipitate the major species present, and the minor constituents will be coprecipitated simultaneously. Both inorganic and organic precipitating agents or collectors are used. For many group separations, the classical inorganic collectors, such as halides, sulfides, hydroxides, and sulfates, are satisfactory. Sometimes simply adjusting the pH of the solution is sufficient. More frequently, it has been found preferable to use one of the many organic precipitating systems as the collector; in this respect, a mixture of

8-quinolinol (also called oxine and 8-hydroxyquinoline), tannic acid, and thionalide (*123–127*) has found wide applicability. Miasoyedova (*128*) has reviewed many different organic systems.

Frequently the concentration of inorganic ions present in the sample is insufficient to form a satisfactory precipitate with which the coprecipitation can occur. In these instances, it is usually necessary to add a carrier ion to provide the bulk of the precipitate. Traditionally iron, aluminum, the rare earths, and manganese have been used as carriers (*91, 129, 130*). Magnesium hydroxide has also been used (*42*). These are, however, subject to several disadvantages. In particular, the heavy metal salts introduce contamination unless they are especially purified. Also, there may be difficulties associated wth removing the excess of the carrier cation, though this can frequently be accomplished by extraction or by a secondary precipitation.

When adding a carrier ion, it is important to evaluate which cation to use and how much of it should be added. The element of choice should have properties similar to those of the constituents being sought, but it should not cause an interference later in the analysis. Minczewski (*3*) provides helpful leads for this. The quantity of carrier element to be added should be sufficiently large that a workable quantity of precipitate forms quickly but not so large that excess impurities are introduced or unwanted ions are coprecipitated in quantity.

Many examples of the application of coprecipitation to the analysis of minor constituents are available. Joyner et al. (*70*) review these for the analysis of trace metals in sea water; Mallory (*131*) obtained quantitative recovery of 14 elements from sea water using thioacetamide with tin and indium as carriers. Seven heavy metals present as impurities in common salts have been separated and preconcentrated using oxine and thionalide with aluminum as the carrier (*132*).

Organic reagents which are insoluble or only sparingly soluble in water and which form insoluble compounds by reaction with the metals of interest are dissolved in a small quantity of a volatile water-miscible organic solvent. This mixture is added to the aqueous sample and the organic solvent is evaporated. As the organic reagent crystallizes, it is enriched in any trace elements which form insoluble compounds with the organic reagent. This technique, called cocrystallization, has been used extensively for the preconcentration of sea water (*70*) using a variety of reagents depending on the elements being analyzed. General purpose reagents have included: 1-nitroso-2-naphthol (*133*), thionalide (*134*), 2-mercaptobenzimidazole (*135*), and potassium rhodizonate (*136*); more specific reagents have included ammonium dipicrylaminate (*137*) and α-benzoinoxime (*138*). Co-crystallization has the added advantage that the precipitate can be taken up directly into as small a portion of organic solvent as possible and aspirated as such into the burner. This provides an extremely useful separation and preconcentration approach for the trace analysis of many elements.

Of less broad interest, but nevertheless of frequent utility, are many specific separation procedures for one or a few elements. Though separation and preconcentration methods for just one or two elements are not usually essential for flame spectrometric methods of analysis, they sometimes are faster or more convenient than the methods previously discussed. For example, Marczenko (*139*) has described a method for palladium utilizing nickel dimethylglyoximate; a cupferron method for titanium using zirconium as the collector has been discussed (*140*). Submicrogram quantities of thorium have been coprecipitated with barium sulfate (*141*); calcium oxalate has been used to coprecipitate strontium in serum and urine analyses (*142*). Applications of fractional precipitation for specific separations appear in the literature (*143*); a related method involving a selective separation and utilizing precipitation by cation release from homogeneous solutions has been described (*144*).

The techniques involved with precipitation, separation of the crystals, and redissolution prior to flame analysis all require special care. Precipitation from homogeneous solution and other standard techniques of gravimetric analysis are recommended. Though normal apparatus can be used effectively, special care must be exercised to avoid contamination and losses. With proper technique, numerous systems have been described which have shown recoveries in the 100% range.

D. ADDITIONAL METHODS

The three methods discussed previously are the ones which have found the broadest use for separation and preconcentration preceding flame spectrometric determination. In this section methods which have been used successfully in specific cases and other methods which are in various stages of development are discussed.

Volatilization methods, either to remove the matrix or for separation of minor constituents from a matrix, have been known for many years and have been discussed at length by a number of authors (*3–5*). Most frequently used is the approach of removing the matrix element by volatilization, leaving behind a concentrated solution of the minor constituents.

This method is likely the oldest known method of concentration and it still works well in many instances. Losses due to adsorption, partial volatilization, or entrainment of the sample must be guarded against; contamination from the air is also a problem. This method is frequently combined with the previously discussed methods to effect a further concentration of the sample by evaporation of the solvent. Numerous examples of this approach are available; examples are in the analysis of fresh water (*70, 145, 146*), and the analysis of impurities in acids (*147*). Various other matrix elements can also be separated by volatilization methods. Silica and silicon-containing samples are treated for removal of the matrix silicon as the tetrafluoride (*148*), and other matrix elements such as arsenic, chromium, germanium, selenium, and tin can be removed by volatilization methods. Minczewski (*3*) includes references to a number of interesting specialized volatilization methods and furnishes tables indicating which matrix elements and which trace elements can be separated by volatility.

Separation and preconcentration of minor constituents by volatilization is considerably more difficult and less utilized than volatilization of the matrix. Adsorption losses, incomplete diffusion of the element in question through the matrix, and incomplete conversion of the element to a volatile form by treatment with some reagent make complete recovery of the minor constituent almost impossible. Contamination is much less of a problem here than with volatilization of the matrix. Many of the same elements can be volatilized at the trace level which could be volatilized as matrix elements; for example, arsenic, germanium, selenium, silicon, tin, and several additional elements such as boron and antimony can also be separated. Most distillations from solution generally involve compounds in which covalent bonds prevail. The distillation of methyl borate (*149*) and of arsine (*150*) are typical examples. Somewhat less typical are the distillation of traces of zinc in high purity aluminum, gallium, and indium and the volatilization of impurities found in molten metals and refractory oxides at high temperatures.

Schuty and Turekian (*151*) have recommended sublimation in preference to the normal methods of evaporating aqueous matrices; they list freedom from spattering, minimal losses due to partial volatilization, and absence of fractional crystallization as the principal advantages. Beyer and Aepli (*152*) have also used this approach to good advantage with ammonium chloride.

Two methods related to volatilization have also been described which are useful for preconcentration but not for separations: cationic concentration by freezing and a thermal preconcentrator for use with flame spectro-metric burners. Malo and Baker (*153*) and Smith and Tasker (*154*) have described freeze concentration for the analysis of trace inorganics in water. Their results indicate that this technique may have significant application, though it has several disadvantages—namely, it is time consuming and some selective ion incorporation in the ice has been observed, which would contribute to losses. Studies incorporating a thermal preconcentrator have been described (*155–157*). With each of these, the intent is to evaporate the solvent, leaving the salt particles in the gas stream, and then discarding the condensed "pure" solvent. Ideally the thermal preconcentrator would remove all of the solvent and would serve as an aspirator with the burner operating in a total consumption manner. Studies thus far have indicated that a relatively high temperature in the evaporation chamber and a cold condensation chamber are preferred. In addition the choice of materials is important as they can effect serious losses and contamination as well as contributing to "memory effects." The results obtained to date with such thermal preconcentrators are highly encouraging. Several of the devices are rapid and exhibit excellent properties: the effects of water entering the flame are largely eliminated, the sensitivity (*158, 159*) using the thermal preconcentrator plus burner was consistently greater than that observed with a standard aspirator-burner combination (*155*). Detection limits, both those measured and those calculated from the sensitivities, were also consistently better with the preconcentrator than without it (*155*). The further development of these integral thermal preconcentration units should provide significant advantages as they permit elimination of specific chemical preconcentration steps.

Various electrochemical techniques for purposes of separation and preconcentration have been excellently reviewed (*4, 5, 160*). These methods are generally time consuming and require several hours or more. With the exception of this disadvantage, these methods would be far more frequently used. The methods depend upon the electrodeposition of metal ions, either as major or minor constituents, at the cathode of an electroly-sis cell. Electrodeposition can be carried out either under constant current, constant voltage, or controlled potential of the working electrode. Controlled potential operation is frequently required if one wishes to make a separation, but for most purposes relative to flame methods this approach is not required. Both liquid mercury and solid metal electrodes, for example, platinum, can be used; however, for most purposes the use of mercury is highly advantageous because of the high overpotential for hydrogen evolution from mercury which permits electrolysis from acid solution. The entire sample present in the form of an amalgam following the electrodeposition can be dissolved and the resultant solution analyzed; however, this means that an extremely high concentration of mercury will be present. Alternatively, the mercury can be distilled off leaving behind the solute materials; the only losses should be due to relatively volatile elements. Another method is anodic stripping wherein a drop of mercury is made the anode and the solute elements are electrolytically dissolved by the application of a controlled potential slightly below that for the dissolution of mercury.

For purposes of concentrating trace impurities in pure metals, the technique of zone refining or melting should be useful. This technique is still relatively undeveloped for these purposes, but it has considerable potential. Shapiro (161) and Konovalov (162) have described the application of zone melting to analytical chemistry by concentrating trace impurities in one end of the test block. Jones and McDuffie (163) have studied the effect of chelating agents in the removal of trace metals from urethane by zone refining. Adsorption or sorption by crystals has also been shown to be a useful means for separating and concentrating trace quantities of metal ions (164–167). For example, this approach has been used for the analysis of cesium.

Using one gram of ammonium 12-molybdophosphate (AMP) microcrystals, the quantitative transfer of rubidium and cesium from one liter of sea water at pH 2 has been reported (70). Feldman and Rains (167) developed a method for the analysis of as little as 6 ng/ml of Cs in water, utilizing its adsorption on AMP in a batch type of operation with aluminum ion used as a flocculant to facilitate collection of the molybdophosphate, followed by dissolution of the molybdophosphate, extraction, and finally measurement by flame spectrometry.

E. Method Selection

The decision as to which separation or preconcentration technique to use can be most time consuming. Occasionally it is possible to find compilations of data and literature references which are immensely helpful. Minczewski (3), for example, has indicated by means of periodic tables which elements can be separated and preconcentrated by a number of different techniques including matrix extraction, trace extraction, coprecipitation, matrix precipitation, and the volatilization of trace or matrix elements. Joyner and co-workers (70) approached the question differently. Using sea water as a common sample, they compiled extensive data on the specific method used, the elements analyzed, the results obtained, and literature references for each element or sample system. For example, six organic reagents are listed for co-crystallization of trace elements and, for fifteen elements, solvent extraction systems of proved or possible use are included. In those instances where compilations or reviews are not available or are out of date, the analyst must refer to the literature or to other sources of information in order to collect the necessary data on which to base a decision. From the total of such data, it becomes much easier to select the best separation and preconcentration technique for a particular analysis under any specific set of conditions.

REFERENCES

1. G. D. Christian, *Anal. Chem.*, **41**, [1] 24A (January 1969).
2. J. A. Dean, *Flame Photometry*, McGraw-Hill, New York, 1960, Part 4.
3. J. Minczewski, in *Trace Characterization, Chemical and Physical* (W. W. Meinke and B. F. Scribner, eds.), U.S. Dept. of Commerce, National Bureau of Standards, Monograph 100, 1965, p. 385.
4. A. Mizuike, in *Trace Analysis, Physical Methods* (G. H. Morrison, ed.), Wiley-Interscience, New York, 1965, p. 103.
5. R. E. Thiers, in *Trace Analysis* (J. H. Yoe and H. J. Koch, eds.), Wiley, New York, 1957, p. 637.
6. Annual Reviews of Analytical Chemistry, *Anal. Chem.*, **38** (1966); **40** (1968).
7. W. Slavin, *Appl. Spectry.*, **20**, 281 (1966).
8. A. B. Calder, *Anal. Chem.*, **36** [9] 25A (August 1964).
9. O. L. Davies, *Design and Analysis of Industrial Experiments*, Hafner, New York, 1963, pp. 99–144.
10. C. L. Grant, *Developments in Applied Spectroscopy* (W. K. Baer, A. J. Perkins, and E. L. Grove, eds.), Vol. 6, Plenum Press, New York, 1968, p. 115.
11. W. J. Youden, *Ind. Eng. Chem.*, **43**, 2059 (1951).
12. W. J. Youden, *Statistical Methods for Chemists*, Wiley, New York, 1951, pp. 33–39.
13. H. J. M. Bowen, *Proc. S. A. C. Conf.*, Nottingham, England, 1965, p. 25.
14. F. J. Flanagan, *Geochim. Cosmochim. Acta*, **31**, 289 (1967).
15. R. E. Thiers, J. F. Williams, and J. H. Yoe, *Anal. Chem.*, **27**, 1725 (1955).
16. C. Blincoe, *Anal. Chem.*, **34**, 715 (1962).
17. D. H. Freeman and E. C. Kuehner, paper presented at the 1st Materials Research Symposium, N.B.S., Gaithersburg, Md., Oct. 1966.
18. D. E. Robertson, *Anal. Chem.*, **40**, 1067 (1968).
19. R. L. Mitchell, *J. Sci. Food Agr.*, **11**, 553 (1960).
20. G. G. Eichholz, A. E. Nagel, and R. B. Hughes, *Anal. Chem.*, **37**, 863 (1965).
21. W. Dyck, *Anal. Chem.*, **40**, 454 (1968).
22. F. K. West, P. W. West, and F. A. Iddings, paper presented at the 1st Materials Research Symposium, N.B.S., Gaithersburg, Md., Oct. 1966.
23. W. J. A. Steyn, *Agr. Food Chem.*, **7**, 344 (1959).

24. H. D. Chapman, *World Crops*, Sept. 1964.
25. E. C. Dunlop, in *Treatise on Analytical Chemistry* (I. M. Kolthoff and P. J. Elving, eds.), Part I, Vol. 2, Wiley-Interscience, New York, 1961, p. 1051.
26. T. T. Gorsuch, *Analyst*, **84**, 135 (1959); **87**, 112 (1962).
27. C. E. Gleit and W. D. Holland, *Anal. Chem.*, **34**, 1454 (1962); *Nature*, **200**, 69 (1963).
28. C. E. Gleit, *Anal. Chem.*, **37**, 314 (1965).
29. C. E. Mulford, *At. Abs. Newsletter*, **5**, 135 (1966).
30. J. E. Shott, Jr., T J. Garland, and R. O. Clark, *Anal. Chem.*, **33**, 506 (1961).
31. W. A. Rowe and K. P. Yates, *Anal. Chem.*, **35**, 368 (1963).
32. H. J. M. Bowen, *Anal. Chem.*, **40**, 969 (1968).
33. S. Fujiwara and H. Narasaki, *Anal. Chem.*, **40**, 2031 (1968).
34. J. E. Burroughs, W. C. Kator, and A. I. Attia, *Anal. Chem.*, **40**, 658 (1968).
35. J. Belisle, C. D. Green, and L. D. Winter, *Anal. Chem.*, **40**, 1006 (1968).
36. N. H. Suhr and C. O. Ingamells, *Anal. Chem.*, **38**, 730 (1966).
37. C. O. Ingamells, *Anal. Chem.*, **38**, 1228 (1966).
38. K. Govindaraju, *Anal. Chem.*, **40**, 24 (1968).
39. B. Bernas, *Anal. Chem.*, **40**, 1682 (1968).
40. F. G. Viets, Jr., *Agr. Food Chem.*, **10**, 174 (1962).
41. A. J. Bedrosian, Thesis, Rutgers, The State Univ., New Brunswick, N.J., 1965.
42. D. C. Burrell, *Anal. Chim. Acta*, **38**, 447 (1967).
43. R. Herrmann and C. T. J. Alkemade, *Chemical Analysis by Flame Photometry*, Wiley-Interscience, New York, 1963.
44. C. E. Mulford, *At. Abs. Newsletter*, **5**, 88 (1966).
45. Yu. A. Zolotov and N. M. Kuz'min, *Z. Anal. Chem.*, **22**, 654 (1967).
46. T. C. Scott, E. D. Roberts, and D. A. Cain, *At. Abs. Newsletter*, **6**, 1 (1967).
47. J. A. Dean, in *Developments in Applied Spectroscopy* (J. E. Forrette and E. Lanterman, eds.), Vol. 3, Plenum Press, New York, 1964, p. 207.
48. J. A. Dean, in *Developments in Applied Spectroscopy* (E. Davis, ed.), Vol. 4, Plenum Press, New York, 1965, p. 443.
49. H. A. Flaschka and A. J. Barnard, *Chelates in Analytical Chemistry*, Marcel Dekker, New York, 1967.
50. T. Higuchi and A. F. Michaelis, *Anal. Chem.*, **40**, 1925 (1968).
51. H. Irving and R. J. P. Williams, in *Treatise on Analytical Chemistry* (I. M. Kolthoff and P. J. Elving, eds.), Part I, Vol. 3, Wiley-Interscience, New York, 1961, p. 1309.
52. G. H. Morrison and H. Freiser, *Solvent Extraction in Analytical Chemistry*, Wiley, New York, 1957.
53. D. D. Perrin, *Organic Complexing Agents*, Wiley-Interscience, New York, 1964.
54. G. K. Schweitzer and W. Van Willis, in *Advances in Analytical Chemistry and Instrumentation* (C. N. Reilley and F. W. McLafferty, eds.), Vol. 5, Wiley-Interscience, New York, 1966, p. 169.
55. J. Stáry, *The Solvent Extraction of Metal Chelates*, Macmillan, New York, 1964.
55a. J. A. Dean, *Chemical Separation Methods*, Van Nostrand Reinhold, New York, 1969.
56. F. Chou and H. Freiser, *Anal. Chem.*, **40**, 34 (1968).
57. J. L. Dick and M. H. Kurbatov, *J. Am. Chem. Soc.*, **76**, 5245 (1954).
58. L. M. Melnick and H. Freiser, *Anal. Chem.*, **27**, 462 (1955).

59. W. J. Blaedel and J. T. Haupert, *Anal. Chem.*, **38**, 1305 (1966).
60. H. Freiser, *Anal. Chem.*, **40** [5] 522R (1968).
61. D. N. Purohit, *Talanta*, **14**, 353 (1967).
62. E. W. Berg and K. P. Reed, *Anal. Chim. Acta*, **36**, 372 (1966).
63. J. E. Allan, *Spectrochim. Acta*, **17**, 459, 467 (1961).
64. J. E. Allan, *Analyst*, **86**, 530 (1961).
65. E. Berman, *At. Abs. Newsletter*, **6**, 57 (1967).
66. R. R. Brooks, B. J. Presley, and I. R. Kaplan, *Anal. Chim. Acta*, **38**, 321 (1967).
67. R. R. Brooks, B. J. Presley, and I. R. Kaplan, *Talanta*, **14**, 809 (1967).
68. M. J. Fishman and M. R. Midgett, in *Trace Inorganics in Water—Advances in Chemistry Series #73* (R. B. Gould, ed.), American Chemical Society, Washington, D.C., 1968, p. 230.
69. D. W. Hessel, *At. Abs. Newsletter*, **7**, 55 (1968).
70. T. Joyner, M. L. Healy, D. Chakravarti, and T. Koyanagi, *Env. Sci. & Tech.*, **1**, 417 (1967).
71. E. Lakanen, *At. Abs. Newsletter*, **5**, 17 (1966).
72. R. E. Mansell and H. W. Emmel, *At. Abs. Newsletter*, **4**, 365 (1965).
73. R. E. Mansell, *At. Abs. Newsletter*, **4**, 276 (1965).
74. M. Montagut-Buscas, J. Obiols, and E. Rodriquez, *At. Abs. Newsletter*, **6**, 61 (1967).
75. W. Slavin, *At. Abs. Newsletter*, **3**, 141 (1964).
76. S. Sprague and W. Slavin, *At. Abs. Newsletter*, **2**, 11 (1964).
77. J. B. Willis, *Anal. Chem.*, **34**, 614 (1962).
78. E. J. Hahn, D. J. Tuma, and J. L. Sullivan, *Anal. Chem.*, **40**, 974 (1968).
79. J. L Jones and R D. Eddy, *Anal. Chim. Acta*, **43**, 165 (1968).
80. G. F. Kirkbright, M. K. Peters, and T. S. West, *Analyst*, **91**, 411 (1966).
81. E. N. Pollock and L. P. Zopatti, *Anal. Chem.*, **37**, 290 (1965).
82. M. Suzuki, M. Yanagisawa, and M. Takeuchi, *Talanta*, **12**, 989 (1965).
83. F. K. West, P. W. West, and T. V. Ramakrishna, *Env. Sci. & Tech.*, **1**, 717 (1967).
84. J. Stáry and J. Smizanska, *Anal. Chim. Acta*, **29**, 545 (1963).
85. J. B. Tyler, *At. Abs. Newsletter*, **6**, 14 (1967).
86. D. E. Green, J. A. B. Haslop, and J. E. Whittey, *Analyst*, **88**, 522 (1963).
87. N. Jordanov and L. Futekov, *Talanta*, **12**, 371 (1965).
88. I. S. Levin, *Talanta*, **14**, 801 (1967).
89. E. Jackwerth, *Z. Anal. Chem.*, **202**, 81 (1964); **206**, 269 (1964); **211**, 254 (1965); **216**, 73 (1966).
90. C. L. Luke, *Anal. Chim. Acta*, **39**, 447 (1967).
91. J. E. Portmann and J. P. Riley, *Anal. Chim. Acta*, **35**, 35 (1966).
92. G. E. Boyd, S. Lindenbaum, and Q. V. Larson, *Inorg. Chem.*, **3**, 1437 (1964).
93. M. H. Campbell, *Anal. Chem.*, **40**, 6 (1968).
94. T. M. Florence and Y. J. Farrar, *Anal. Chem.*, **40**, 1200 (1968).
95. H. Green, *Talanta*, **11**, 1561 (1964).
96. T. Groenewald, *Anal. Chem.*, **40**, 863 (1968).
97. J. R. Knapp, R. E. Van Aman, and J. H. Kanzelmeyer, *Anal. Chem.*, **34**, 1374 (1962).
98. See Ref. (*3*), p. 403.
99. G. H. Faye, *Anal. Chem.*, **37**, 696 (1965).

100. T. R. Sweet and H. W. Parlett, *Anal. Chem.*, **40**, 1885 (1968).

101. G. F. Kirkbright, A. M. Smith, and T. S. West, *Analyst*, **92**, 411 (1967).

102. F. Helfferich, *Ion Exchange*, McGraw-Hill, New York, 1962.

103. K. A. Kraus, in *Trace Analysis* (J. H. Yoe and H. J. Koch, Jr., eds.), Wiley, New York, 1957, Chapt. 2.

104. O. Samuelson, *Ion Exchange Separations in Analytical Chemistry*, Wiley, New York, 1953.

105. H. F. Walton, in *Chromatography* (E. Heftmann, ed.), Reinhold, New York, 1961.

106. K. Govindaraju, *Anal. Chem.*, **40**, 24 (1968).

107. F. W. E. Strelow, J. H. J. Coetzee, and C. R. Vanzyl, *Anal. Chem.*, **40**, 196 (1968).

108. F. W. E. Strelow, W. J. Louw, and C. H. W. Weinhart, *Anal. Chem.*, **40**, 2021 (1968).

109. F. W. E. Strelow, *Anal. Chem.*, **40**, 928 (1968).

110. N. R. Andersen and D. N. Hume, *Anal. Chim. Acta*, **40**, 207 (1968).

111. H. D. Perdue, A. Conover, N. Sanley, and R. Anderson, *Anal. Chem.*, **40**, 1773 (1968).

112. A. Lewandowski and W. Szczepaniak, *Z. Anal. Chem.*, **202**, 321 (1964).

113. R. Hering et al., *J. Prakt. Chem.*, **32**, 291 (1966); **34**, 69 (1966).

114. R. R. Ruch, F. Tera, and G. H. Morrison, *Anal. Chem.*, **36**, 2311 (1964).

115. F. Tera, R. R. Ruch, and G. H. Morrison, *Anal. Chem.*, **37**, 358, 1565 (1965).

116. W. J. Campbell, E. F. Spano, and T. E. Green, *Anal. Chem.*, **38**, 987 (1966).

117. E. F. Spano and T. E. Green, *Anal. Chem.*, **38**, 1341 (1966).

118. G. L. Hubbard and T. E. Green, *Anal. Chem.*, **38**, 428 (1966).

119. D. G. Beichler, *Anal. Chem.*, **37**, 1054 (1965).

120. J. Korkisch, *Sep. Sci.*, **1**, 159 (1966).

121. J. Korkisch and K. A. Orlandini, *Anal. Chem.*, **40**, 1952 (1968).

122. T. Y. Toribara and P. S. Chen, Jr., *Anal. Chem.*, **24**, 539 (1952).

123. M. C. Farquar, J. A. Hill, and M. M. English, *Anal. Chem.*, **38**, 208 (1966).

124. R. L. Mitchell, *Commonwealth Bur. Soil Sci. Techn. Commun.* No. 44, 1948.

125. R. L. Mitchell and R. O. Scott, *Spectrochim. Acta*, **3**, 367 (1948).

126. R. L. Mitchell and R. O. Scott, *Appl. Spectry.*, **11**, 6 (1957).

127. E. F. Cruft and J. Husler, *Anal. Chem.*, **41**, 175 (1969).

128. G. V. Miasoyedova, *Zh. Anal. Khim.*, **21**, 598 (1966).

129. G. F. Reynolds and F. S. Tyler, *Analyst*, **89**, 579 (1964).

130. R. Ko and P. Anderson, *Anal. Chem.*, **41**, 177 (1969).

131. E. C. Mallory, Jr., in *Trace Inorganics in Water-Advances in Chemistry Series No. 73* (R. B. Gould, ed.), American Chemical Society, Washington, D.C., 1968, p. 281.

132. R. L. Dehm, W. G. Dunn, and E. R. Loder, *Anal. Chem.*, **33**, 607 (1961).

133. H. V. Weiss and M. G. Lai, *J. Marine Res.*, **18**, 185 (1960); *Anal. Chim. Acta*, **25**, 550 (1961).

134. M. G. Lai and H. V. Weiss, *Anal. Chem.*, **34**, 1012 (1960).

135. H. V. Weiss and M. G. Lai, *Anal. Chim. Acta*, **28**, 242 (1963).

136. H. V. Weiss and M. G. Lai, *Anal. Chem.*, **32**, 475 (1960).

137. H. V. Weiss and M. G. Lai, *J. Inorg. Nucl. Chem.*, **17**, 366 (1961).

138. H. V. Weiss and M. G. Lai, *Talanta*, **8**, 72 (1961).

139. Z. Marczenko, *Chim. Anal. (Paris)*, **46**, 286 (1964).

140. J. O. Hibbits, J. Kallmann, W. Giustetti, and H. K. Oberthin, *Talanta*, **11**, 1462 (1964).

141. C. W. Sill and C. P. Willis, *Anal. Chem.*, **36**, 622 (1964).

142. J. Jordan, *At. Abs. Newsletter*, **7**, 48 (1968).

143. F. H. Firsching et al., *Anal. Chem.*, **40**, 152 (1968).

144. P. F. S. Cartwright, *Anal. Chem.*, **40**, 1157 (1968).

145. E. A. Boettner and F. I. Grunder, in *Trace Inorganics in Water—Advances in Chemistry Series No. 73* (R. B. Gould, ed.), American Chemical Society, Washington, D.C., 1968, p. 236.

146. M. D. Kleinkopf, *Bull. Geol. Soc. Am.*, **71**, 1231 (1960).

147. J. H. Oldfield and E. P. Bridge, *Analyst*, **85**, 97 (1960).

148. W. T. Rees, *Analyst*, **87**, 202 (1962).

149. M. Freegarde and J. Cartwright, *Analyst*, **87**, 214 (1962).

150. W. T. Oliver and H. S. Funnell, *Anal. Chem.*, **31**, 259 (1959).

151. D. F. Schuty and K. K. Turekian, *Geochim. Cosmochim. Acta*, **29**, 259 (1965).

152. K. W. Beyer and O. T. Aepli, *Anal. Chem.*, **29**, 1779 (1957).

153. B. A. Malo and R. A. Baker, in *Trace Inorganics in Water—Advances in Chemistry Series No. 73* (R. B. Gould, ed.), American Chemical Society, Washington, D.C., 1968, p. 149.

154. G. H. Smith and M. P. Tasker, *Anal. Chim. Acta*, **33**, 559 (1965).

155. D. W. Ellis and D. R. Demers, paper presented at the Eastern Analytical Symposium, Nov. 1968, New York.

156. C. Veillon and M. Margoshes, paper presented at the XIII Colloquium Spectroscopicum Internationale, Ottawa, Canada, June 1967.

157. J. P. Misland and S. Elchuck, *At. Abs. Newsletter*, **7**, 71 (1968).

158. J. Mandel and R. D. Stiehler, *J. Res. Natl. Bur. Stds.*, **53**, 155 (1954).

159. R. K. Skogerboe, A. T. Heybey, and G. H. Morrison, *Anal. Chem.*, **38**, 1821 (1966).

160. J. J. Lingane, *Electroanalytical Chemistry*, 2nd ed., Wiley-Interscience, New York, 1958, p. 416.

161. J. Shapiro, *Science*, **33**, 2063 (1961).

162. E. E. Konovalov et al., *Zh. Anal. Khim.*, **18**, 624 1500 (1963).

163. L. N. Jones and B. McDuffie, *Anal. Chem.*, **41**, 65 (1969).

164. J. Krtil and V. Kourin, *J. Inorg. Nucl. Chem.*, **25**, 1069, 1191 (1963).

165. J. van R. Smit, J. J. Jacobs, and W. Robb, *J. Inorg. Nucl. Chem.*, **12**, 104 (1960).

166. J. van R. Smit, W. Robb, and J. J. Jacobs, *Nucleonics*, **17**, 116 (1959).

167. C. Feldman and T. C. Rains, *Anal. Chem.*, **36**, 405 (1964).

11   Trace Analysis and Micromethods

Roland Herrmann

DEPARTMENT OF MEDICAL PHYSICS
UNIVERSITY OF GIESSEN
GIESSEN, WEST GERMANY

I. Introduction

In general, the topic will be limited to conventional atomic emission or atomic absorption methods using normal chemical flames. Nonflame methods, treated in Chapter 4, will be mentioned only if they offer substantial advantages. It will be assumed that the sample is already in solution form as a result of appropriate prior treatment (Chapter 10). Special precautions in sample preparation will be mentioned later.

In trace analysis the concentration of test element will be in the neighbor-hood of the detection limit but the amount of sample available will be between 1 and 10 ml, that is, sample volumes adequate for nebulization directly. By contrast, microanalysis refers to a situation in which solution quantities are of the order of 0.1 ml down to a nanoliter in size, although the concentration of test element may be 20 times the detection limit or higher.

The boundary between the two categories is not sharp. If, in trace analysis, there is sufficient sample available, the concentration can be increased by appropriate prior concentration, such as partial evaporation of the solvent or removal by ion exchange followed by elution in a limited volume of solvent. Likewise, if the element concentration is quite high in microanalysis, a suitable dilution will convert the sample into a normal concentration and volume of material that can be handled in a conven-tional manner.

Extreme care must be exercised during all operations and handling so that no impurities or contaminants are introduced, and that no loss occurs due to volatilization, adsorption on container surfaces, or the like. It is essential that a blank be carried through the entire procedure. The advan-tage of having a sensitive method is lost if the sample is contaminated without a means for correction.

II. Trace Analysis

In trace analysis the instrumental conditions must be optimized to obtain the lowest detection limit and reasonable results despite low concentrations of test element (1–4). Although it is conceivable that the concentration level could be raised by preliminary chemical separation methods, it is sometimes desirable to alter the apparatus to improve the detection level. The latter approach represents a single expenditure of effort as contrasted with repetitive laboratory operations required for a concentration step. Instrument optimization to attain the smallest detect-able concentration levels has been treated extensively in other chapters.

The accuracy and precision requirements of trace analysis (and micro-methods) are generally not as stringent as the requirements of macro-analysis. Indirect methods are quite suitable for trace analysis contrary to prejudice against such methods.

Detection limits in atomic emission depends substantially on the excitation energy required for the specific element and the energy actually available in the excitation source. The method depends on the magnitude of the background and its fluctuation, which contributes a noise component to the net emission signal. For atomic absorption the significant factors are the product of the oscillator strength of the absorbing atom and the statistical weight factor, often called simply the "*gf* factor", plus the fluctuations arising during nebulization and from the light source.

A. EQUIPMENT

Equipment cost is not a criterion of suitability. Frequently, a single-beam, filter photometer will be adequate for detecting strongly emitting elements by atomic emission. Cost and resolving power often go hand in hand (Chapter 5). Resolution need not be unusually high when there exists a dearth of emission lines from either the flame in atomic emission or from the source in atomic absorption. A double-beam spectrometer compensates better for fluctuations arising in the source and thus impairing detection limits through its noise component. In deciding between different commercial instruments, one should be chary in accepting the detection limits quoted by the vendor. Different definitions for detection limits, whether these limits are applicable to actual samples or only pure standards, and whether these limits apply to average instruments, are factors to consider. Although often optimistic, detection limits quoted by the vendor can be used as a first approximation, lacking a definitive listing such as will appear in Volume 3.

In trace analysis the line-to-background ratio becomes the criterion for distinguishing the presence or absence of an element; the disappearance of the signal into the background fluctuations restricts the detection limit (unless a differential mode is employed, Section III.D). One attempts to adjust parameters, such as bandpass, to achieve as large a ratio of line to background as possible. Usually, emission line intensities decrease proportionally when spectral bandpass is decreased, whereas flame background (due to molecular emitters) decreases quadratically, at least until the critical slit width is reached. In atomic absorption, the detection limit usually is determined by the ratio of the source half-band width to the half-band width of the resonance line involved. Consequently, adequate detection limits can often be obtained with equipment of lesser resolution than can be done in the emission mode. Light-gathering power of the monochromator becomes important when working with emission signals which are of very low luminous density. Use of interference filters, where applicable, drastically improves the total flux reaching the photodetector. The latter must be chosen to provide maximum response with a minimum noise factor. Aging inevitably occurs, and periodically the detector should ·be compared against a new unit.

Experimental skill plays a large part in the ability of an operator to achieve the lowest possible detection limits. This is particularly true for equipment in which a large number of options are available, such as flame type, observation height, bandpass, wavelength selected, fuel gases, amplifier-detector settings, scale expansion, and noise suppression. Equipment does deteriorate in time. Mirrors become dirty and perhaps tarnished. Nebulizer characteristics change with usage, burner orifices become constricted with carbon deposits. Lamp output changes with time.

B. Flames

Both atomic absorption and atomic emission methods require efficient breakdown of the sample into an atomic vapor. Elements which tend to form stable hydroxides or oxides will suffer impairment in their detection limits. For these elements, special flames are frequently required; reducing flames for elements forming refractory oxides, and flames low in atomic hydrogen for elements forming MOH species.

No single flame will provide optimal detection limits for all elements and matrix compositions. Atomic emission methods are preferred for the alkali metals. Molecular emission methods with "cold" and often shielded flames are recommended for nonmetals such as phosphorus and sulfur. By contrast, atomic absorption and atomic fluorescence give better detection limits for elements such as Cd, Zn, and Mg. A laboratory engaged in trace analysis must have emission, absorption, and fluorescence capabilities.

For easily excited elements, such as the alkalis, low temperature flames are quite satisfactory and minimize ionization interferences. Detection limits for sodium are often so good with these flames and simple equipment that the limiting factors are reagent purity, impurities introduced by diluents, and from particulate matter in air. When the laboratory air is contaminated with the test element, such as would occur from alkali-containing dust or spray from the seashore swept into the surrounding countryside, practical detection limits are lowered. Equipment should be housed inside a laminar flow hood in these situations. Some help is given by a protective gas sheath surrounding the flame. The problem is alleviated somewhat when the test element is emitting its desired radiation from the inner conal gases, or when a separated flame is employed. In these cases, external room air does not penetrate into the emission area.

The mixing ratio of oxidant to fuel must be carefully chosen for many elements. For elements which tend to form oxygen compounds, strongly reducing hydrocarbon flames, or a nitrous oxide–acetylene flame, are recommended. The latter flame possesses a real advantage in trace and microanalysis because the relatively slow rise rate of the combustion gases in the flame means a longer residence time for atoms within the area of observation and thus increased emission and absorption sensitivity.

When faced with elements that form extremely refractory compounds or which are difficult to excite, radio-frequency plasmas may offer a remedy (5, 6). The radiation temperatures attainable lie between 8000 to 16,000°C. Furthermore, the discharge takes place only at the inner surface; consequently, curvature of calibration curves due to self-absorption is less frequent.

. The concentration of ground state atoms and excited atoms in the different zones of a flame can vary due to the differences in temperature and composition of the zones. For the best detection limits, it is important that the profile of the emission or absorption signal be obtained or known so that the region of maximum emission or absorption for each element can be used (Chapter 8 of Volume 1).

The magnitude of the emission or absorption signals is not the sole criterion in evaluating sensitivity. As pointed out in Chapter 13 of Volume 1, it is their magnitude in relation to the signal fluctuation that determines the ultimate detection limit. Therefore, any reduction in flame noise and signal fluctuations will necessarily improve the detection limit. Since the background radiation affects the entire flame, and since the flame in absorption is long and narrow, the characteristic absorption of flame molecules, and scattering by particles within the flame mantle, play a significant role. These factors are especially critical in the far ultraviolet where the resonance lines of arsenic and selenium are located. Argon (entrained air)–hydrogen flames lessen the absorption by flame gases (7, 8).

C. NEBULIZER

The nebulizer should be carefully selected and maintained (see Chapter 3 of this volume). Transfer efficiency directly affects sensitivity as it determines the actual amount of sample, in the form of the aerosol–gas mixture, that reaches the burner. Losses by condensation within the spray chamber detract from the ultimate signal. Although transfer efficiency is improved by heating the spray chamber, there is no corresponding improvement in the detection limit unless smaller droplets are formed at the same time. In the flame large drops vaporize more slowly, possibly incompletely, thus leading to scattering of light and, equally significantly, to signal loss due to insufficient time for atomization of salt particles within the light path of the spectrometer. Scattering problems were discussed in Chapter 10 of Volume 1.

The application of ultrasonic nebulizers to flame emission and atomic absorption offer the advantages of increased sensitivity and lower sample consumption, often as little as 0.2 ml total sample volume (Chapter 3, Section II.D.1).

D. Sample Preparation

If the detection limits cannot be improved by altering the operational procedure to permit a direct determination of the test element, it may be possible to increase the concentration of test element by suitable chemical separation methods (Chapter 10). When resorting to prior separation methods, special consideration should be given to methods which simultaneously remove interferents while concentrating the test element. Extraction methods are to be recommended. Due to the specificity of atomic absorption, nonspecific extraction systems can be used. Extraction not only is faster than other separation methods, but substantial enhancement of emission and absorption signals can be obtained using organic solvents (9). The 3- to 5-fold improvement in atomic absorption is due largely to a smaller drop-size distribution in the aerosol. The manyfold increase in many emission signals is due to effects such as a change in flame temperature and chemiexcitation phenomenon.

Filtration is of little value to remove particulate matter since the filter material can contribute impurities to dilute solutions or analyte can be removed through adsorption onto the filter surfaces. Centrifugation is much better and is applicable whenever substantial differences in density exist between solvent and suspended particles. Ashing may serve to concentrate the sample. Although simple to carry out, wet ashing suffers the disadvantage that impurities are introduced through the acids used and losses may occur through spattering. Dry ashing in a muffle furnace offers less danger from contamination so long as the sample is covered, but easily vaporized elements will be lost. Ashing at about 100°C in a vacuum (1–10 Torr) with introduction of a stream of ozone (*10*) or in a stream of active oxygen in a rf field of 13.56 MHz (*10a*) circumvents loss of elements with low boiling points but is more complex in execution.

E. Sources of Error

When working with low concentrations of test element, one has to be constantly on the alert for sources of contamination or losses by adsorption. Impurities can be leached from the walls of the containing vessels unless proper chemical and physical preparation of the containers precedes their use. For example, alkalis and calcium in glass containers can be leached out. One remedy is to coat the interior of containers with silane preparations, followed by a thorough rinsing cycle with deionized water. Use of polyethylene vessels is perhaps the best remedy, but even here trouble can develop, particularly with silver solutions. Frequently overlooked as a source of contamination is the stopper. When opening or closing a container, glass grinds against glass releasing alkalis and calcium. Substitution of plastic stoppers removes this source of difficulty.

Standard solutions are usually prepared from concentrated stock solutions, prepared as described in Chapter 13, and stored in polyethylene bottles. Dilutions should be prepared at frequent intervals, perhaps daily. They can undergo a loss in strength due to adsorption onto the container walls or as a consequence of hydrolysis. Thus, slightly acidic solutions are preferred and are more stable than neutral or basic solutions. However it is difficult to generalize, since the effect will vary considerably with the particular element and chemical composition of the solution. Solutions much weaker than 100 μg/ml should be prepared daily. At extremely low concentrations, purity of the solvent can present a problem. Common contaminants must be guarded against; many distilled water supplies have detectable levels of metal impurities. Freshly distilled water that has been passed through a mixed-bed ion exchanger is recommended. Microbes and algae will grow in standard solutions which are kept for a long period of time. These foreign bodies may remove a portion of certain elements.

III. Microanalysis

Microanalytical methods are important in biomedical problems, air pollution studies, and so on. Sample volumes are usually very small.

Consequently, it is not feasible to wait for the quasiequilibrium state to be reached in a flame. Sample volumes are simply insufficient for normal flame emission or atomic absorption work. Moreover, in the final measurement step, sensitivity losses are compounded through dilution by the flame gases. Special sample introduction techniques are required and rapid measuring methods must be used. Often an integrating mode is obligatory to measure the transient signals normally obtained (see Chapter 6, Section V.C).

A. Sample Introduction of Nanoliter Volumes

Apparently there is no commercial pipetting equipment to handle nanoliter-size samples. Consequently, a homemade pipet and pipetting

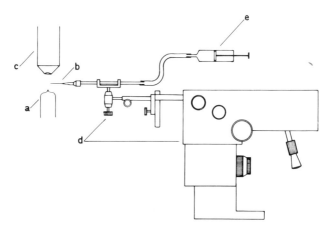

Fig. 1. Micropipetter and platinum wire inserter: a. Wire for insertion into flame; b. capillary micropipet; c. microscope to observe operations; d. micromanipulator; and e. syringe.

technique will be briefly described with the help of Fig. 1. Begin with a glass capillary of narrow internal diameter. Heat in a luminous flame of a Bunsen burner and, when the glass softens, draw out both ends in jerks and remove the capillary from the flame. After examination under a microscope, select a portion of the drawnout capillary that possesses a suitable inner diameter. Cut the capillary to give a normal capillary with a long tip (Fig. 1a). The thin measuring capillary is inserted by its thick end into a micromanipulator which allows motion of the capillary in all directions. With the aid of a microscope, place a calibration mark on the thin capillary. The actual position of the calibration mark is not critical as long as the same capillary is used for all solutions, although the volume designated by the mark should not differ from the nominal value by orders of magnitude. Directions for pipetting are in Ref. (*11*).

Because the ratio of surface area to volume is large with small drops, evaporation effects are important and must be suppressed by surrounding the solution to be measured with a lighter solvent, such as liquid paraffin

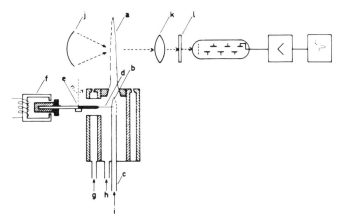

Fig. 2. Schematic diagram of flame emission system for nanoliter-size samples: a. Outer mantle of flame; b. inner cone; c. capillary; d. center point; e. platinum wire; f. magnet; g. outer oxygen supply; h. oxygen; i. fuel; j. mirror; k. lens; and l. filter (*18*).

or petroleum ether. Of course, none of the paraffin should enter the pipet, but the liquid interface is readily discernible with practice.

Rather than attempting to spray nanoliter-size samples into a flame, they are deposited on a platinum wire (Fig. 2e) by means of the micro-pipetter. The solvent is allowed to evaporate at room temperature. Any auxiliary heating must be at a low temperature so that the solvent evaporates quietly without boiling or spattering.

The dried sample on the wire must be introduced reproducibly into the flame whose composition, size, and shape must remain constant. The same location for insertion must be used for all samples and standards, and the sample must be inserted at a constant velocity. Usually the samples are introduced at the upper edge of the inner combustion zone because a reducing atmosphere with relatively high temperature exists at this location. Solute vaporization will occur quickly, and metal oxides are more likely to be reduced. The thinness of this zone creates a mechanical problem in controlling the movement of the sample and holder. One method involves a platinum wire which is moved around a vertical axis by magnets. The terminal point of motion is determined by a mechanical stop. A spring pulls back the empty platinum wire. Applications are discussed in Refs. (*12, 13*).

B. HELIUM-GLOW PHOTOMETER (*13*)

The physical layout of the helium-glow photometer is sketched in Fig. 3. Operation of the photometer is based on the fact that electrical excitation

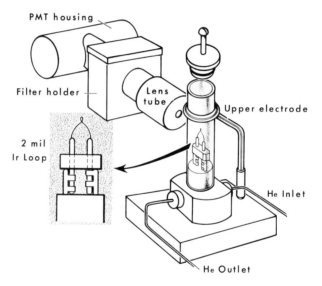

Fig. 3. Helium-glow photometer (*13*).

of helium at atmospheric pressure by the application of a high voltage radio-frequency electric field generates very energetic, metastable helium atoms that can transfer their energy to impurity atoms and excite them to emit their characteristic radiation.

Two to 10 nl volumes of sample, containing 10^{-13} to 10^{-11} moles of the element to be determined, are placed in the loop of iridium wire. These quantities are about 10^{-4} as much material as required for flame emission or atomic absorption. The loop forms one part of the rf electrode pair, and is within a quartz tube which can be purged with helium. The other part of the rf electrode pair is a ring of wire around the outside of the tube, about one tube diameter above the apex of the iridium wire. After the inflowing helium (3 ml/sec) has purged the atmospheric gases from the chamber (about 75 sec), the rf field (27 MHz) is applied, and the iridium wire is heated electrically. The sample is excited and emits its spectra as a brief flash lasting 0.1 to 0.3 sec as it is volatilized into the helium glow which surrounds the wire apex. A portion of the emitted light is collected by a collimating lens and is transmitted through appropriate light filters to a photomultiplier tube. The anode current is integrated; the magnitude of the integral is proportional to the amount of material in the sample.

C. OTHER VAPORIZATION METHODS

The inability of flames to completely reduce an analytical sample to free atoms has greatly reduced the sensitivity of flame methods for many elements. Moreover, since flame emission and atomic absorption generally operate with a solution, the volume of solvent introduced during sample preparation lowers the analyte concentration and thereby the sensitivity.

Graphite furnaces at elevated temperatures provide a reducing environment favorable for the existence of free atoms. Approximately 0.1-mg samples are evaporated in a 2000–3000°C graphite tube. Inconvenience of sample prehandling has detracted from the method. This and the need for reduced pressure when sputtering atoms from a metallic cathode in cool and hot hollow cathodes has handicapped the use of sputtering methods. As pointed out in Chapter 5, where these methods are described in detail, sensitivity achieved may be as low as picograms for one percent absorption.

Another method for solid samples consists of controlled burning of a blend of solid propellant powder (oxidant and fuel mixture) and the pulverized material under study (14). Detection limits lie in the part-per-million range. The method requires considerable sample preparation, especially for materials that are not easily pulverized.

Laser evaporation of material from a preselected area of an analytical specimen into the optical path of an atomic absorption spectrometer also yields detection limits on the part-per-million level for some elements, but they are highly dependent on the host material (see Chapter 4, Section III.B).

D. Repetitive Optical Scanning in the Derivative Mode (15)

The method minimizes the need for a monochromator of high resolving power. A quartz plate made to vibrate at 145 Hz is mounted behind the entrance slit of the monochromator. The ac amplifier is synchronized with the oscillations of the quartz plate. When the amplifier is tuned to twice the frequency of vibration, the second derivative of the spectrum is obtained as the wavelength is scanned. This mode of operation for flame emission spectrometry permits the measurement of weak line spectra without the interference from background radiation or broad band spectra. The signal-to-noise ratio is improved by eliminating the low frequency flicker noise of the flame. The elimination of interferences from bands and flame structure provides an improvement in detection limits of most elements in the presence of many matrix ions. Typical values of detection limits range from 0.02 ng/ml for lithium to 3.0 ng/ml for calcium. An analysis can be performed with as little as 50 μl of solution, which makes it applicable to microanalysis.

E. Special Methods for Mercury

Analyses for mercury have a special place. Special arrangements for vaporization and dissociation are unnecessary since mercury has such a high vapor pressure at room temperature that it can be measured directly in normal atmospheres (16–18) by atomic absorption. From its inorganic compounds mercury has been amalgamated onto a copper spiral; upon heating the spiral the mercury vapors are carried by an air current into the optical path of the spectrometer. After wet digestion, mercury in urine (19) is reduced to the metallic state by tin(II) and volatilized into an air stream; detection limit is 0.2 ng in 1 ml of urine.

REFERENCES

1. R. Herrmann and C. T. J. Alkemade, *Chemical Analysis by Flame Photometry*, Interscience-Wiley, New York, 1963.
2. G. H. Morrison (ed.), *Trace Analysis, Physical Methods*, Wiley, New York, 1965.
3. W. W. Meinke and B. F. Scribner (eds.), *Trace Characterization, Chemical and Physical*, National Bureau of Standards Monograph 100, U.S. Government Printing Office, Washington, D.C., 1967.
4. E. Pungor, K. Toth, and J. Konkoly-Thege, *Z. Anal. Chem.*, **200**, 321 (1964).
5. R. Mavrodineanu and R. C. Hughes, *Spectrochim. Acta*, **19**, 1309 (1963).
6. R. H. Wendt and V. A. Fassel, *Anal. Chem.*, **38**, 337 (1966).
7. D. W. Ellis and D. R. Demers, *Anal. Chem.*, **38**, 1943 (1966).

8. K. Zacha and J. D. Winefordner, *Anal. Chem.*, **38**, 1537 (1966).
9. J. A. Dean, *Flame Photometry*, McGraw-Hill, New York, 1960.
10. F. Dittel, *Z. Anal. Chem.*, **228**, 432 (1967).
10a. C. E. Gleit, *Am. J. Med. Electronics*, **2**, 112 (1963).
11. D. J. Prager, R. L. Bowman, and G. G. Vurek, *Science*, **147**, 606 (1965).
12. D. Stamm and R. Herrmann, *Z. Klin. Chem.*, **3**, 193 (1965).
13. G. G. Vurek and R. L. Bowman, *Science*, **149**, 448 (1965).
14. A. A. Venghiattis, *At. Abs. Newsletter*, **6**, 19 (1967).
15. W. Snelleman, T. C. Rains, K. W. Yee, H. D. Cook, and O. Menis, *Anal. Chem.*, **42**, 394 (1970).
16. H. Brandenberger and H. Bader, *Chimia*, **21**, 597 (1967).
17. M. B. Jacobs, L. J. Goldwater, and H. Gilbert, *Am. Ind. Hyg. Assoc. J.*, **22**, 276 (1961).
18. P. Müller, *Exp. Cell. Res. Suppl.*, **5**, 118 (1958).
19. G. Lindstedt, *Analyst*, **95**, 264 (1970).

12   Evaluation of Data

W. G. Schrenk

CHEMISTRY DEPARTMENT
KANSAS STATE UNIVERSITY
MANHATTAN, KANSAS

I. Introduction

Flame spectrometry, whether it is emission, atomic absorption, or fluorescence, is used primarily as an analytical technique for elemental analysis. The useful analytical range varies with the method, element under consideration, and the spectral line used. Concentrations of many elements can be successfully determined to the part-per-billion (nanogram/milliliter, ng/ml) range.

Many other factors also are involved in analytical flame spectrometry. These include rate of sample atomization, fuels and oxidants used, as well as their ratios, nebulizer-burner design, sensitivity of electronic equipment, use of organic solvents, hollow cathode lamp intensity (for atomic absorption), region or area of the flame that is viewed by the monochromator, and the presence of extraneous elements in the sample being analyzed.

Since extraneous substances in the sample may influence intensity measurements, it is necessary for the analyst to be aware of such factors. These have been discussed in Chapters 10–12 of Volume 1 of this series and reference to that information should be made if questions arise concerning such interference effects.

This chapter will discuss the procedures involved in quantitative evaluation of data in general terms, as well as the special problems associated with each of the techniques. Methods that can be used to eliminate or minimize interference effects also will be discussed.

II. General Considerations

A. Selection of Spectral Line

Spectral lines for emission purposes can be selected by reference to tables of line sensitivities. One of the best sources of information are the tables compiled by Meggers et al. (1). These tables refer primarily to arc and spark emission spectroscopy, but nevertheless those lines listed as most sensitive also are most sensitive for flame emission. Flame excitation is a low energy excitation, thus most of the lines of low sensitivity are not seen in flame emission. Other useful references include Dean (2), Mavrodineanu and Boiteux (3), Herrmann and Alkemade (4), and this series of volumes.

Margoshes (5) has published the criteria to be used in predicting useful lines for atomic absorption spectrometry. Robinson (6) has tabulated useful lines for a large number of elements and Ramirez-Muñoz (7) has tabulated useful lines for some 68 elements, as has Slavin (8).

B. GENERAL PRECAUTIONS

A number of precautions must be followed routinely in any analytical scheme. These become even more important when the procedure is being used at trace levels of determination, as is common in flame methods.

Instrumental parameters must be optimized for each element, and instrument stability also must be achieved. One must adjust fuel and oxidant flow rates to predetermined values, optical alignment must remain fixed, and electronic instruments need to be "warmed up" until stable.

Glassware and storage containers for stock solutions and standards must meet rigid standards of cleanliness. Dilute solutions require special care. Contamination from dust particles, smoke, cleaning solutions, etc., should be avoided. This is especially true of elements such as sodium, potassium, and calcium. Stock solutions for preparing standards can be stored reasonable lengths of time, but glass containers should not be used for alkali or alkaline earth metal standards. Polyethylene bottles are very useful for such solutions. Soft glass should be avoided in all cases.

There have been some reports of contamination and erratic data resulting from fuel and oxidant fuel lines. If this occurs it is wise to place filters in the lines. Filters of loosely packed glass wool seem effective if they are changed as necessary. Particular attention to this factor is necessary if acetylene and compressed air are being used.

III. Standards and Working Curves

A. STANDARD SOLUTIONS

Substances used as standards should be nonhygroscopic, spectrally pure, reagent grade chemicals. Some standards may be best prepared from pure metals dissolved in reagent quality hydrochloric acid. Quantitative analysis texts can be consulted for recommended chemicals for standards. Most companies selling atomic absorption and flame emission instruments also sell chemicals for standards and some have stock solutions available already prepared. Stock solutions should be very carefully prepared, using precautions already given, and should have a concentration greater than any used to establish working curves. For example, stock solutions are commonly prepared at concentrations of 1000 to 2000 μg/ml. Proper dilution of stock solutions are then made to prepare working standards. *The final accuracy of any flame method depends on the accuracy of the comparison standards*, therefore, exercise the utmost care in their preparation. Directions for many standard solutions will be found in Chapter 13.

B. Preparation of Working Standards

To establish an analytical working curve the simplest procedure is to prepare a series of standards covering the concentration range to be studied. Standards should be prepared by proper dilution of the stock solution. The same solvent system should be used for the standards as is used for the samples. Where possible, the anion associated with the standard should be the same as that present in the sample. For example, if the sample is placed in solution in 1 N HCl the standard also should contain 1 N HCl. Avoid, if possible, di- or trivalent anions, since they have more affect on calibration curves than do most monovalent anions.

The standard solutions thus prepared are simple solutions involving the cation under test, the associated anion, and a solvent similar to the unknowns. Such standards are useful if no interferences of other ions present in the unknown occur, and under such circumstances can serve very effectively to determine concentrations of the unknown. If interferences do occur in the unknown, measurements based on such simple standards may introduce error into the analysis. In such cases further sample preparation or working standards is necessary. The methods used in these cases will be discussed in a later section of this chapter.

C. Establishment of the Working Curve—Flame Emission

Data obtained with the working standard solutions can be plotted in various ways depending, in part, on the type of instrumentation in use. If a graphical (chart paper) readout system is in use, data such as that shown in Fig. 1 will be obtained. It may be noted that no background correction is necessary in this case. Thus a simple measure of intensity is possible where peak height (in millimeters) is a measure of the signal intensity for each standard.

These data may be plotted as intensity vs concentration, as shown in Fig. 2, and, in this case, a linear calibration curve is obtained. From such a curve it is a simple matter to determine the concentration of an unknown. Also since the curve is linear a simple algebraic solution may be

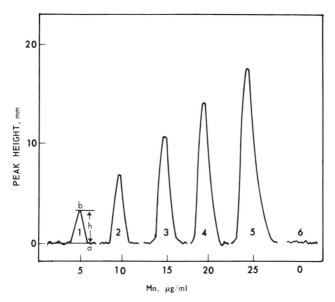

Fig. 1. Flame emission signals of manganese standards (4032 Å).

established. In this example the slope of the calibration curve may be calculated and the relation,

Concentration of unknown = slope × peak height, in millimeters   (1)

can serve to determine the concentration of the unknown.

In the use of such a calibration curve or algebraic equation on a day-to-day basis it is essential to check the curve every day. Minor fluctuations of fuel and oxidant pressures or minor changes in sample flow rates may change the slope of the calibration a slight amount. To compensate for such changes it is necessary to use one of the higher standards and return the instrument to the same peak height by slight changes in fuel and/or oxidant pressures.

The possibility of instrument drift or other changes occurring during a series of determinations also should be checked by running a standard after several unknowns. Such a procedure will detect any minor variations during any particular set of determinations. Between each analytical determination a blank or distilled water should be run to prevent any possible "build-up" of sample in the burner-aspirator parts and to ensure return of the signal to the "zero" level.

If the instrumentation used does not permit graphical presentation of the data, visual readings obviously will be required. Depending on

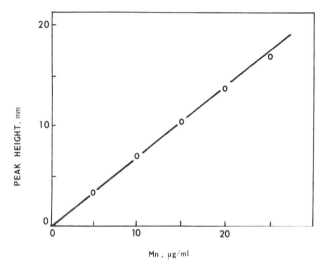

Fig. 2. Concentration of manganese in μg/ml vs peak height in mm.

instrumentation this may be a galvanometer deflection, potentiometer reading, digital readout, or percent transmission scale. The problem one faces is to obtain a measure of the relative emission intensity. To obtain such results two measurements usually are needed. The type of readings required are illustrated in Fig. 1. A reading at the wavelength of the maximum signal (*b*) and of the background (*a*) are required; the difference is the signal intensity of standard 1. It is essential that all readings of maximum intensity be at the same wavelength. It is preferable to carefully choose the wavelength for (*b*) and also (*a*) and use the same settings for each reading. If the background is constant and low it may not be necessary to measure (*a*) for each sample, but only repeat readings for (*a*) at predetermined intervals, for example, every five samples.

1. *Effect of Varying Background Signal*

Occasionally it is necessary to obtain line emission intensities in a region of variable background. The background signal may arise from several causes. Our concern analytically deals only with proper corrections for such effects. Figure 3 illustrates this effect. The analytical line is rubidium superimposed on the "tail" of a potassium line. Note the rapidly changing background in each case. Graphically the problem can be solved as indicated by reference to Fig. 3. A tangent is drawn as shown in sample 4

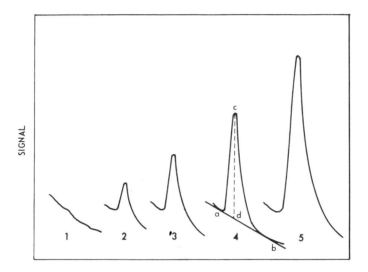

Fig. 3. Rubidium emission signals (7800 Å) in the presence of potassium.

from *a* to *b* to closely approximate the slope of the blank. A vertical line from the dotted tangent, *ab*, is drawn from the point of maximum signal intensity, *c*, to the tangent at *d*. The "true" signal intensity is measured by *cd*. A working curve can then be constructed as was done in Fig. 2.

If scanning is not possible because of the type of instrumentation, the same result can be achieved by making three measurements for each determination. Readings should be taken at wavelengths corresponding to *a*, *b*, and *c* of Fig. 3. Readings *a* and *b* should be equally spaced with reference to *c*. The readings for *a* and *b* are averaged, and the average is subtracted from *c* to give a measure of the signal intensity of rubidium. Obviously it is necessary to exercise extreme care to assure that the three readings are taken at precisely the same wavelengths for each determination.

D. ESTABLISHMENT OF THE WORKING CURVE—ATOMIC ABSORPTION

In atomic absorption the attenuation of a signal from the source, usually a hollow cathode, is a measure of the concentration of the element in the sample solution. A hollow-cathode lamp of the element being determined is required. It has been shown that the attenuation of the signal obeys Beer's law. That is

$$\log \frac{I}{I_0} = -kdC \tag{2}$$

where I_0 is the original intensity, I is the intensity after passing through the sample, d is the path length through the sample, and C is the concentration. In atomic absorption I_0 is the intensity of the particular emission line in use as emitted by the hollow-cathode lamp, I is the intensity, of the same line, after passage through the flame, and d is the thickness of the flame. Since $\log(I_0/I) = A$, the absorbance, then, $A = kdC$ where k is the absorptivity and has a characteristic value for each system studied. Since d is constant one can see that A is directly and linearly related to C for each system considered, with the slope of the calibration curve being determined by k and d.

The above considerations suggest that the best way to plot atomic absorption working curves is to use absorbance versus concentration. A common method to determine absorbance relates this quantity to percent transmission. If the blank is adjusted to 100% transmission the sample will produce a reading of something less than 100%. Under these circumstances the absorbance A is

$$A = 2 - \log(\text{percent transmission}) \tag{3}$$

where 2 is the log of 100 and the second term to the right of the equality sign is the log of the percent transmission of the standard or unknown.

If the atomic absorption instrument in use has a visual meter for signal intensity measurements the usual procedure is to adjust the meter for full scale (100% transmission) with the hollow cathode at operating current and a blank (solvent) being aspirated into the flame. With no energy entering the monochromator the meter should read zero. The monochromator must have previously been set at the proper wavelength for the desired signal.

After the adjustment to 100% transmission on the blank, a series of standards covering the appropriate range are successively aspirated and percent transmission values obtained. These data are converted to absorbance values using Eq. (3). A table of such values simplifies the calculations. Semilog paper also can be used if desired. Table 1 presents a set of such data. A straight-line calibration curve should result, as is indicated by the data. Concentrations of unknown solutions can then be determined by reference to the graph or by use of the equation $A = mC$, where A is

TABLE 1

ATOMIC ABSORPTION CALIBRATION DATA FOR ZINC

Zn, μg/ml	Percent absorption	Percent transmission	Absorbance $(2 - \log \%T)$
0.50	6.1	93.9	0.027
1.00	11.1	88.9	0.051
2.00	21.6	78.4	0.106
4.00	36.9	63.1	0.201
8.00	59.3	40.7	0.390

the absorbance, m is the slope of the curve, and C is the desired concentration.

The usual precautions concerning checking of instrument drift periodically and cleanliness of the aspirator-burner combination should, of course, be followed.

If a graphical "readout" system is available with the atomic absorption instrument in use, results such as those shown in Fig. 4 may be obtained. In this figure the base line (a) represents the signal from the hollow cathode passing through a blank. Signal 6 represents an approximately 30% absorption signal placed on the calibration sheet for use if data is to be treated as in Table 1. Curves 1, 2, 3, 4, and 5 represent decreases in signal caused by the presence of zinc in varying amounts. The signal, in percent, may be calculated from a working curve as shown in Fig. 5. For example, signal 1, produced by 8 μg/ml of zinc, represents 59.3% absorption. Such a representation of data is often simpler to use than it is to adjust the blank to a fixed position on the recorder chart and no signal to another fixed point. This representation also allows for "scale expansion," a technique involving electronic amplification of weak signals for ease of reading but which places "zero" signal off the chart.

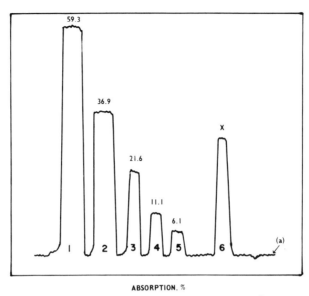

Fig. 4. Atomic absorption signals of zinc standards (2138.6 Å) and unknown, *X*.

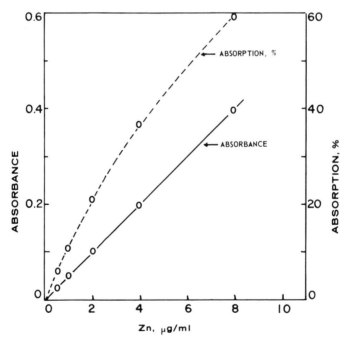

Fig. 5. Calibration curves of atomic absorption signals of zinc using absorbance and percent absorption vs zinc concentration in µg/ml.

312

Data obtained from Fig. 4 can be treated by conversion to absorbance by use of the relation $A = 2 - \log \%T$ (where $\%T = 100 - \%$ absorption). These data can then be plotted as shown in Fig. 5. The absorbance versus concentration is linear, as predicted by Beer's equation. For comparison purposes percent absorption versus concentration also is presented. Note the curvature of the percent absorption plot of the analytical data.

E. NONLINEAR CALIBRATION CURVES

There will be frequent occurrences of nonlinearity in working curves. Such curves can be used for analytical purposes but generally reduce

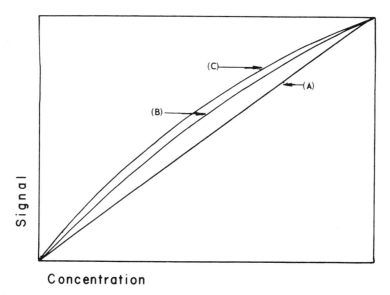

Fig. 6. Calibration curves for emission signals of magnesium (2852 Å): (A) 0–10 μg/ml; (B) 0–50 μg/ml; (C) 0–100 μg/ml.

accuracy slightly. A common type of nonlinearity occurs when higher concentrations of standards are used. In this case "self-absorption" of signals in the outer, cooler portions of the flame cause departure from linearity. This effect is illustrated in Fig. 6. Here we see three calibration curves relating emission intensity of magnesium (2852 Å) to concentration. Curves were normalized at maximum intensity for comparison purposes.

The calibration curve, 0–10 μg/ml, is linear; 0–50 and 0–100 μg/ml show nonlinearity in varying degrees.

Two approaches are possible. If a calibration curve is nonlinear it is advisable to increase the number of standard solutions to accurately determine the shape of the curve. This will tend to reduce the error in subsequent analyses based on the curve. A second approach is to dilute the test solution so determinations may be made using lower concentrations, thereby placing the sample in a linear portion of a calibration curve. Care needs to be exercised in either case. If a nonlinear working curve is used it is essential to read concentration vs signal very carefully. A mathematical expression for a nonlinear curve is usually too complicated for routine use. If a dilution technique is used care must be exercised with respect to accuracy of the dilution. Contamination must be avoided by using clean glassware and solvents, volumes of solutions must be determined accurately, and careful records must be kept on all steps so final calculations relating to the original sample concentration can be determined.

IV. Other Evaluation Techniques

A. Sample Bracketing

If only a few samples are to be determined, a sample bracketing technique can provide excellent analytical results. After an initial run on the unknown, one chooses two standards (concentrations corresponding to A and B of Fig. 7), one with a concentration lower and the other higher than that of the unknown. If the calibration curve in the region used is linear (as AXB of Fig. 7) one can then use the following expression to obtain the concentration of the sample:

$$\text{Conc(sample)} = \text{Conc } A + \left[\frac{\text{Signal(sample)} - \text{Signal}(A)}{\text{Signal}(B) - \text{Signal}(A)} \right]$$
$$\times (\text{Conc } B - \text{Conc } A) \quad (4)$$

If the working curve is nonlinear in the region used, the same technique may be used, but one must be careful to select standards to produce as small an interval as possible. Inspection of Fig. 7 will illustrate why. Assume $ACDB$ represents the working curve. If the bracketing standards are of concentrations A and B, use of the mathematical expression will yield a concentration of X when Z is correct. If the interval between standards is reduced to C and D the error will be the difference between Y and Z. The method assumes that the part of the working curve between the two standards can be treated as a straight line.

If the working curve is linear, Eq. (4) can be reduced to

$$\text{Conc(sample)} = \text{Conc } A + k(\text{Conc } B \text{ Conc } A) \tag{5}$$

where k is the slope of the calibration curve and is numerically equal to the quantity in brackets in Eq. (4).

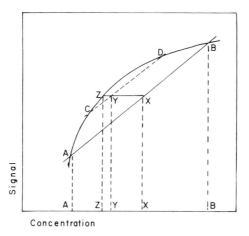

Fig. 7. Sample bracketing technique with linear and nonlinear calibration curves.

The method of sample bracketing is capable of producing very accurate concentration determinations or determinations when only a few samples are to be run.

B. METHOD OF STANDARD ADDITIONS

The standard addition technique involves the use of the unknown and the unknown plus the addition of a known amount of standard. Frequently it is advisable to add amounts of known in several different concentrations and make a series such as, X, $X + A$, $X + 2A$, and $X + 3A$, where X is the unknown and A represents a definite amount of known.

If the simple case is used, that is X and $X + A$, the amount of X can be obtained by simple subtraction of A from the measured concentration of

$X + A$. The technique is especially useful for very low concentrations of X where calibration data may be doubtful or subject to considerable percentage error. The method is useful only if calibration curves are linear and pass through the origin.

A simple graphical technique may be employed to determine unknown concentrations and is especially useful if several standard addition samples are measured. This technique is illustrated in Fig. 8. The signal produced

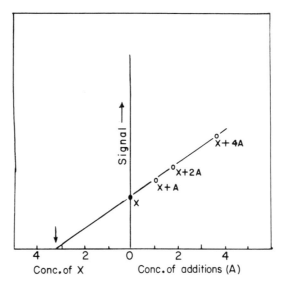

Fig. 8. Standard addition method for the determination of the concentration of an unknown.

by the sample is on the vertical signal axis; samples with standard additions are to the right. The linear curve is then extrapolated to the sample concentration axis to the left of the signal axis. The figure indicates an unknown containing 3.3 units of concentration. Concentration units may be in any convenient dimension, obviously units of addition and unknown must be the same and graphically presented similarly.

C. DILUTION TECHNIQUES

Dilution methods are useful, especially for nonroutine samples. Gilbert et al. (9) have suggested the procedure for flame emission, but it is equally adaptable to atomic absorption procedures. The apparent concentration of the sample is compared with another aliquot diluted to produce a sample with one-half the concentration of the first. Standards may be compared similarly. If the original and the diluted sample, as read off the calibration curve, agree closely the data may be assumed correct. Further dilutions may be useful to improve accuracy if the concentration does not become too small to give a meaningful signal. The principal advantage of the technique in nonroutine cases is that serious deviation in apparent concentration from the 1:2 ratio indicates the presence of interferences. If such deviation does occur, other methods or treatments of the sample will be necessary to provide accurate analytical data.

D. INTERNAL STANDARDIZATION

The principle of internal standardization is well known and widely used in arc and spark emission spectroscopy. Dean (2) has discussed the usefulness of the technique for flame emission. The method can compensate for minor variations in physical properties of solutions and may be used to minimize variation in feed rates, changes in droplet size, and fluctuations in pressure of the fuel and oxidant gases.

A constant amount of an internal standard ion, not present in the original sample, is added to each working standard and to each unknown solution sample. In flame emission the two signals, from the test element and the internal standard, are measured, preferably simultaneously. If simultaneous measurements are not possible, they should be made successively in as short a time interval as possible. The calibration curve is a plot of the intensity ratio of element/standard versus concentration of the test element. Such a plot on log-log paper should give a straight line.

A suitable internal standard element should (1) have a useful spectral line at a wavelength close to that of the test element, (2) be a line that arises from an energy transition similar to that of the test element, (3) have an ionization potential similar to that of the test element, and (4) not be present in the test solution as a contaminant.

The usefulness of this technique is limited to relatively simple systems and should not be relied on to compensate for major interference effects, since the test element intensity and that of the internal standard may not be affected similarly by the interfering substance(s).

The internal standard has merit however in compensating for minor variations caused by changes in physical properties such as viscosity, density, and surface tension, all of which affect sample uptake rate and drop size. It is useful also as a convenient check to determine if operating conditions have remained constant during the course of a series of determinations.

V. Interference Effects on Analytical Data

As with most analytical techniques, interferences to the processes used to obtain quantitative data can and do occur. Many classifications of interference effects have been suggested. It is convenient to classify them as spectral, physical, and chemical. It is not the intention of this chapter to detail interference effects since this has been done in Chapters 9–12 of Volume 1 of this series of books. A brief review is however useful to establish background for the analytical approaches to interferences that are described in the following sections of this chapter.

A. SPECTRAL INTERFERENCES

In flame emission spectrometry spectral interferences are quite common. This type of interference can occur due to the proximity of a test element line and an emission line of an extraneous element in the test solution. For example, the copper line at 3247 Å and a manganese line at approximately the same wavelength will spectrally interfere with one another. Numerous cases of such interferences are known. The analyst therefore should check wavelength tables to determine if this possibility may exist in the system under study.

Molecular band emission may occur from components of the fuel and oxidant gases. Strong bands due to OH, CH, CN, and C_2 are observed in some types of flames. For example, OH bands interfere with copper emission. Band emission also may occur due to elements in the sample. If calcium is present, bands due to CaO or CaOH may be present.

Still another source of spectral interference is a general increase in background due to the solvent system being used.

It has been said that spectral interferences do not occur in atomic absorption spectroscopy. However Koirtyohann and Pickett (10) presented evidence of band spectra interference, and Fassel et al. (11) have shown that spectral line interferences can be a problem. The work of Frank et al. (12) also indicates that spectral line interferences can occur.

It is true, nevertheless, that spectral interferences are much less important in atomic absorption than in flame emission.

B. PHYSICAL INTERFERENCES

Physical interferences are those caused by the physical properties of the test solutions. They may cause the behavior of the solution to change as the sample is being aspirated into the flame and include such properties as viscosity, density, vapor pressure, surface tension, temperature, solvents, high salt concentrations, and possibly colloidal particles.

Signal intensity, for example, will be depressed by any property that decreases the sample flow rate. As the sample enters the flame, droplets can cause light to be scattered and thus contribute to the general background. Acid concentrations also cause general changes of signal intensity.

Usually physical interferences are not great on a percentage basis. They may, however, be sufficiently large to cause serious problems with the accuracy of analytical results, especially at low concentration levels.

C. CHEMICAL INTERFERENCES

Interferences can arise in flame emission and atomic absorption spectroscopy due to chemical interaction of species present in the flame, including sample, solvent, fuel, and oxidants. There are several common kinds of chemical interferences that the analyst should be aware exist.

1. Ionization Effects

Consider a system containing an easily ionized element such as lithium in the presence of a second easily ionized element potassium. The emission or absorption signals of both lithium and potassium are enhanced in the presence of the other. The introduction of lithium in a sample of potassium depresses the ionization of potassium and thus increases the concentration of the ground state or excited atoms of potassium. This type of signal enhancement is called ionization interference. The effect can be quite large, as illustrated in Fig. 9.

2. Chemical Effects

Many examples exist of the formation of chemical compounds and radicals in a flame that affect the ground state or excited state populations of the test element. Perhaps the most studied system is the effect of the phosphate anion on calcium. The signal intensity of calcium is decreased severely by the presence of the phosphate ion. This effect has been ascribed to compound formation in the flame, thus calcium is removed as a ground state atom and signal intensity decreased. Numerous other similar cases exist but will not be discussed here.

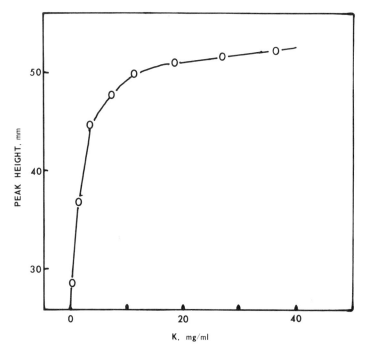

Fig. 9. Effect of progressive levels of potassium on the [background corrected] emission signal of rubidium. (Reproduced from (*13*) by courtesy of the American Chemical Society.)

A second chemical effect results from the reaction of oxygen and/or hydrogen in the flame with the test element. Elements such as calcium and magnesium form very stable oxides that usually are not decomposed in the flame. Hydroxides also may be formed, and their presence in flames has been confirmed.

A third effect which may be chemical in nature is that caused by samples of high salt concentrations. This effect is largely caused by a change in the

signal intensity distribution in the flame. A redistribution of signal intensities can occur which appears in analytical data as an interference effect.

D. DETECTION OF INTERFERENCE EFFECTS

Detection of the existence of interferences is necessary to development of
any analytical procedure in flame spectrometry. One procedure that can
be used has already been referred to. Gilbert et al. (9) have suggested that
the test solution be prepared at two concentration levels, preferably a
2:1 ratio. If interferences are present the concentration ratio, as determined
from a calibration curve, will deviate from the expected 2:1 ratio.

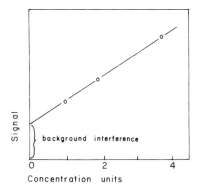

Fig. 10. Detection of background interference in flame spectrometry.

The method of standard addition also provides information concerning
the existence of interference effects. In this case two solutions are prepared,
an unknown and the same unknown plus the addition of a known amount
of standard. If simple subtraction of the concentration of the standard
does not produce a value in close agreement with that obtained by direct
measurement it is evidence that interferences are occurring.

General background interference may be detected easily by a graphical
approach, as illustrated in Fig. 10. If the unknown is prepared at two or
three concentration levels, for example at concentrations A, $2A$, and $4A$,
and signal readings obtained a graphical representation of results may
appear as in Fig. 10. If no interference is present the curve should pass
through the origin. The background interference, in terms of measured
signal, is indicated in the figure.

VI. Methods for Elimination or Reduction of Interferences

If preliminary experiments indicate the existence of interferences in the system under investigation, the analyst is faced with devising some technique or procedure to either compensate for or eliminate these interferences. There are a number of approaches to the problem, and they depend, in part, on the nature of the interference encountered.

One general approach to eliminate interference effects involves modification of the working standard solutions used to prepare the calibration curves. Instead of using a simple, single component standard solution one may approximate the composition of the unknown and include all elements in the standard solution that are part of the unknown. For example, suppose one wishes to determine copper in plant tissue. Standard solutions containing fixed amounts of all major constituents of plant ash could be placed in the working standards along with the varying amounts of copper. Standards of this type have been used for years in arc and spark emission spectroscopy. Such a method also provides an excellent system for addition of an internal standard if that is desirable. The assumption made in this method is that the substances added to the standard solutions affect the working curve in the same manner and to the same degree as in the unknown solution. The method has the disadvantage of requiring tedious preparation. New standards are required for each type of sample.

A. REMOVAL OF INTERFERING SUBSTANCES

Another general approach to the problem of interferences is to remove them from the test solution. Various separation methods—extraction, ion exchange, and precipitation—may be used, as discussed in Chapter 10.

B. IONIZATION INTERFERENCES

Ionization interferences can be severe when easily ionized elements are present in the test solution and an easily ionized element is under analysis. This type of interference produces enhancement effects that can be substantial. For example, in flame emission the signal from potassium can be increased as much as 50% by addition of lithium or cesium to the test solution. In atomic absorption the signal enhancement is not as great but may be 10 to 15%. This effect is due to shift in the equilibrium

$$M° \rightleftharpoons M^+ + e^-$$

produced by increasing the number of free electrons in the flame from a second easily ionized species. The equilibrium is shifted to the left, increasing the population of [M°], thereby increasing the absorption signal arising from [M°] and the emission signal from [M*] produced by excitation of the ground state atom.

A simple procedure to stabilize this effect is to add to the test solution an excess of an easily ionized element. The result of the addition is to essentially "flood" the flame with excess electrons, thus providing stability (and enhancement) to the flame. This effect has been illustrated in Fig. 9 (13). An additional advantage is increased sensitivity. This method is a special case of a "buffer" system as an effective device to control and stabilize conditions in the flame. It is especially useful for determinations of alkali metals in the presence of other alkali metals.

C. DILUTION PROCEDURES

Interference effects can frequently be decreased by dilution of the sample with solvent. Such a procedure is possible only if the element of interest is present in amounts to make dilution possible. Not all interferences respond to such a procedure however. It is necessary therefore to use one of the procedures mentioned earlier to determine if, in fact, the interference effects have decreased. The method is not useful as a procedure to minimize ionization interferences. It is most useful for interferences caused by the presence of extraneous anions or other species that may react chemically with the element under test. It also is effective to minimize self-absorption effects.

D. BUFFERING PROCEDURES

Buffering procedures have been used for many years in arc and spark emission spectroscopy. A special case of buffering was mentioned previously, that of stabilizing the enhancing effect produced by ionization. A buffer may be defined as any added substance that reduces or stabilizes an interference. Addition of interfering substances to a solution may produce a stabilizing effect on the signal produced by the test element. For example, it is possible to add phosphate ion to a solution containing calcium until the calcium emission is no longer affected by the presence of the phosphate ion. The same principle may be applied to interferences caused by the sulfate ion. An excess of an interfering species, if present in all samples, tends to decrease variations between samples. Usually buffering of this type decreases the signal intensity of the test element and therefore is not useful if concentrations are near the detection limit.

E. RELEASING AGENTS

Releasing agents are substances added to the test solution that "release" the test element from the effects of the interfering substances. A well-known case is the use of lanthanum chloride to "release" calcium in flame spectrometry (14). The effect of aluminum on magnesium can be reduced by additions of calcium or strontium chloride (15). The blank solution should contain the releasing agent in the same concentration as the unknown.

Chelating agents may be considered as special cases of releasing agents, since they have a tendency to reduce the effects of the interference. EDTA has been used as a chelating agent with both calcium and magnesium (16–18). The mechanism of action of a chelating agent is considered to be a process that prevents the test element from reacting with another species in the flame. Rains et al. (19) found that phosphate interference to calcium could be minimized by a variety of chelating agents, including glycerol, ethylene glycerol, dextrose, and lactose.

F. INSTRUMENTAL PARAMETERS AND INTERFERENCES

Instrumental factors and operating conditions have an effect on the magnitude and nature of certain interferences in flame spectrometry. Some of these factors can be controlled and others cannot.

Spectral interferences are at least in part related to instrumentation. A spectrometer of high dispersive power can resolve spectral lines close in wavelength while such resolution may not be possible with another spectrometer, as was discussed in Chapter 5.

No single unit of the instrumentation required for flame spectrometry comes in a wider variety of types than does the nebulizer and burner, discussed in Chapter 3. Interferences, especially chemical interferences, differ in magnitude between a laminar flow burner and a turbulent nebulizer-burner. These differences arise partly from the types of fuel and oxidant employed, but also from the nature of the transfer path.

As discussed in Chapter 8 of Volume 1, certain areas of the flame are more efficient than others for atomic absorption and emission processes. Some interferences can be reduced or eliminated by proper choice of the flame area under observation; the area of maximum signal sensitivity and the area of minimum interference are not necessarily the same. Furthermore, the optimum height for emission is usually not the same height as for maximum atomic absorption. Thus, in preliminary work it is advisable to locate the areas of maximum response and minimum interference.

REFERENCES

1. W. F. Meggers, C. H. Corliss, and B. F. Scribner, *Tables of Spectral-line Intensities*, NBS Monograph 32, U.S. Government Printing Office, Washington, D.C., 1961.
2. J. A. Dean, *Flame Photometry*, McGraw-Hill, New York, 1960.
3. R. Mavrodineanu and H. Boiteux, *Flame Spectroscopy*, Wiley, New York, 1965.
4. R. Herrmann and C. T. J. Alkemade, *Chemical Analysis by Flame Photometry*, Interscience-Wiley, New York, 1963.
5. M. Margoshes, *Anal. Chem.*, **39**, 1093 (1967).
6. J. W. Robinson, *Atomic Absorption Spectroscopy*, Marcel Dekker, New York, 1966.
7. J. Ramirez-Muñoz, *Atomic Absorption Spectroscopy*, Elsevier, Amsterdam, 1968.
8. W. Slavin, *Atomic Absorption Spectroscopy*, Interscience-Wiley, New York, 1968.
9. P. T. Gilbert, Jr., R. C. Hawes, and A. O. Beckman, *Anal. Chem.*, **22**, 772 (1950).
10. S. R. Koirtyohann and E. E. Pickett, *Anal. Chem.*, **38**, 585 (1966).
11. V. A. Fassel, J. O. Rasmuson, and T. G. Cowley, *Spectrochim. Acta*, **23B**, 579 (1968).
12. C. W. Frank, C. E. Meloan, and W. G. Schrenk, *Anal. Chem.*, **38**, 1005 (1966).
13. T. E. Shellenberger, R. E. Pyke, D. B. Parrish, and W. G. Schrenk, *Anal. Chem.*, **32**, 210 (1960).
14. C. H. Williams, *Anal. Chim. Acta*, **22**, 163 (1960).
15. A. C. Menzies, *Anal. Chem.*, **32**, 898 (1960).
16. R. Herrmann and W. Lang, *Z. Ges. Exptl. Med.*, **135** 569 (1962).
17. J. B. Willis, *Nature*, **186**, 249 (1960).
18. J. B. Willis, *Spectrochim. Acta*, **16**, 273 (1960).
19. T. C. Rains, H. E. Zittel, and M. Ferguson, *Talanta*, **10**, 367 (1963).

13 Standard Solutions for Flame Spectrometry

John A. Dean   *and*   *Theodore C. Rains*

DEPARTMENT OF CHEMISTRY
UNIVERSITY OF TENNESSEE
KNOXVILLE, TENNESSEE

ANALYTICAL CHEMISTRY DIVISION
NATIONAL BUREAU OF STANDARDS
WASHINGTON, D.C.

I. Introduction

In all evaluation methods one must have available standard samples and high-purity materials for each element being determined, often in the same matrix as the sample. Quite often several standards must be prepared for a given element, one for each matrix or system that contains the test element. To assist the individual chemist who must prepare these standard solutions, this chapter will outline the preparation of various individual stock solutions and suggest sources where high-purity materials may be obtained or from whom standard samples may be purchased. An excellent report on available standard reference materials and high-purity materials has been compiled by Michaelis (*1*).

A. Sources of Standard Reference Materials

The National Bureau of Standards (U.S.) offers a wide selection of standard alloys, steels, ores, glasses, and miscellaneous types of materials in the form of chips, turnings, powders, or granules. Full information concerning the values that are certified, the weights, and prices is contained in Ref. (2).

The Bureau of Analysed Samples, Ltd. (Newham Hall, Middlesborough, Yorkshire, England) also offers a wide selection of standard samples. Many of these standards may be obtained through the Jarrell-Ash Division of Fisher Scientific Company (590 Lincoln Street, Waltham, Mass. 02154) or from Spex Industries, Inc. (3880 Park Avenue, Metuchen, N.J. 08841).

A series of 20 ceramic powders are supplied by Applied Research Laboratories (9545 Wentworth Street, Sunland, Calif. 91040) for the analysis of barium, calcium, magnesium, and strontium titanates, and titanium dioxide. The standards have a variation of composition to cover the range of impurity constituents normally found in ceramic materials.

In addition to the various standard samples, reference samples are also available. Generally compositions are less accurately known. Smith and Underwood (1023 Troy Court, Troy, Mich. 48084) supply over 1000 different analyzed materials, primarily for student use in universities and colleges.

B. Types of Standard Reference Materials Available

Three aluminum-base alloys prepared in the form of turnings are available from the National Bureau of Standards (NBS), and seven chip samples are issued by the Bureau of Analysed Samples (BAS). These samples possess certificates of analysis for Cu, Fe, Mg, Mn, Pb, Si, Ti, and Zn in all cases; individual samples may contain also Cr, Ga, Ni, Sb, Sn, and V.

Two cobalt-base alloys are issued by NBS and contain Ni, Mn, Si, Fe, Cr, Mo, W, Nb, and Cu.

Nine copper-base alloys are available from NBS and five from BAS, all in chip form. In addition to Cu, Zn, Sn, Pb, Ni, and Fe, certain standards also contain Al, As, Ag, Co, Mn, Sb, Si, S, P, and Mg. Six standards in the form of turnings from high-purity copper are available from Bundensanstalt für Materialprüfung (Berlin-Dahlem, Unter den

Eichen 87, Berlin, Germany). These standards contain less than 0.07% of Pb, Sn, Bi, As, Sb, Fe, and Ni.

Over 120 standards for iron-base alloys are available from NBS and BAS in the form of chips. These include cast iron, ingot iron, B.O.H. and A.O.H. steels, plain-carbon steels, mild steels, and alloy steels. The mild steels and alloy steels contain accurately determined traces of a number of residual elements (W, Co, Ti, As, Sn, Al, Zr, Pb, Nb, Ta, B, and Ag) in addition to the usual constituents: Mn, P, S, Si, Cu, Ni, Cr, V, and Mo.

Two lead-base standards are issued by NBS and one by BAS; these contain Sn, Sb, Bi, Cu, As, and traces of Ag and Ni.

One magnesium-base alloy is available from NBS and one from BAS. The former contains Al, Zn, Mn, Si, Cu, Pb, Fe, and Ni; the latter contains total rare earths, Zn, Zr, Cu plus traces of Mn, Ni, and Fe.

The NBS issues three primary standards of nickel oxide (powder form) and four nickel-base alloys (Monel, nickel-base casting alloy, Ni-Cr alloy, Waspaloy). The BAS supplies one standard (Nimonec 90).

One tin-base alloy is available from NBS and from BAS. The standard contains Sb, Cu, As, Pb and traces of Bi, Ag, Ni, and Fe.

Only one zinc-base alloy is available from NBS. It contains Al, Cu, Mg, and traces of Fe, Mn, Pb, Ni, Sn, and Cd.

One zirconium-base alloy, Zircaloy 2, is issued by NBS. It contains Sn, Fe, Cr, and traces of Ni, Mn, and Cu.

Steel-making alloy standards from NBS and BAS include refined silicon, ferrovanadium, ferrochromium, spiegeleisen, ferrophosphorus, ferroboron, and calcium molybdate.

A set of six cement standards is available from the NBS with these constituents certified: SiO_2, Al_2O_3, Fe_2O_3, TiO_2, P_2O_5, CaO, SrO, MgO, SO_3, Mn_2O_3, Na_2O and K_2O.

Miscellaneous standard samples issued by the NBS include: three alumina refractories, two silica refractories, one chrome refractory, a bauxite, a plastic clay, a glass sand, four glasses, and argillaceous limestone, a silica brick, a burned magnesite, titanium dioxide, and silicon carbide. These are supplemented by three basic slags, a silica brick, a firebrick, and a sillimanite from the BAS.

Among ore samples available are six iron ores from the BAS. A phosphate rock, a Mesabi iron ore, three lithium ores (Spodumene, Petalite, and Lepidolite), one manganese ore, two tin ores, and one zinc ore are available from the NBS. Except for the phosphate rock and Mesabi iron ore, the NBS ores are certified only for the component of interest.

A syenite rock sample and a sulfide ore are also available (Department of Geological Sciences, McGill University, Montreal, P. Q., Canada).

Four horticultural standards of apple, cherry, citrus, and peach leaves have been prepared (Horticultural Department, Michigan State University, East Lansing, Mich.).

II. Aqueous Stock Solutions

High-purity materials (99.9% or greater) can be obtained from several sources: Alpha Metals, Inc. (60 Water Street, Jersey City, N.J.), Johnson Matthey & Co., Ltd. (Hatton Garden, London, E.C.1, England), Spex Industries, Inc. (Section I.A.), and various chemical supply houses, for example, Baker & Adamson, J. T. Baker & Co., Fisher Scientific Co., Mallinckrodt Chemical Works, and Merck and Co. A comprehensive listing of suppliers of high-purity materials for virtually every element is available (1).

In selecting a suitable compound or material for use as a standard, several criteria must be met: (1) It must be available commercially in a high degree of purity and meet A.C.S. specifications or a similar "official" certification. Established methods should be available for confirming its purity (3). (2) The compound should not be hygroscopic, nor should it be efflorescent. It should be stable under the conditions specified for drying. (3) The composition of the selected material must be known; if not, the final stock solution will have to be standardized. (4) Metallic and other contaminants present in these standard materials must not be in amounts greater than that of the trace element to be determined in any particular sample, otherwise mixed standards could not be prepared.

Procedures which provide standard stock solutions containing 1000 $\mu g/ml$ of the element of interest are listed in Appendix 1. Precautions outlined in Chapters 10 and 11 (Section II.E) should be reviewed. Molarity of the reagents used is as follows:

Reagent	Molarity	Reagent	Molarity
Acetic acid	17.4	HNO_3 (69%)	15.4
HCl (38%)	12.0	H_3PO_4 (85%)	14.7
$HClO_4$ (70%)	11.6	H_2SO_4 (94%)	17.6
HF (45%)	25.7	Ammonia (27%)	14.3

Many stock solutions should be slightly acidic to prevent hydrolysis. Unless stated otherwise, final dilution should be made to 1000 ml with deionized water—ordinary distilled water which has been passed through a mixed-bed ion-exchange resin such as Amberlite MB-3, then freshly boiled and cooled to room temperature. Solutions are stored in polyethylene containers unless a special containment material is specified. A crystal of thymol in each stock solution will inhibit bacterial growth.

A. RELATION OF CONCENTRATION UNITS

The interrelationship between various expressions for concentrations is:

1 μg/ml $\equiv$ 1 mg/liter $\equiv$ 1 ppm (w/v)
1 meq/liter = (weight in grams per equivalent/1000) in 1 liter
1 mg atom per liter = (atomic weight in grams/1000) in 1 liter
1 μg/ml = (1 meq/liter)(1000/weight in grams per equivalent)

III. Specialized Stock Solutions

Solutions for clinical, agronomic, and certain other special situations are listed in Appendix 2. These solutions are prepared in the same manner as described in Section II.

IV. Standards for Trace Elements in Petroleum Products

Stable, oil-soluble compounds are available from the National Bureau of Standards (U.S.) as certified standards for 24 elements in petroleum products. Appendix 3 lists the compounds; all are solids. Directions for the preparation of individual compounds have been published (5) for those who may wish to prepare their own materials. Some of the compounds are fairly hygroscopic. As a precautionary measure, bottles containing the standard material should be stored, tightly closed, in desiccators containing P_2O_5, and that, before use, the required amount of a compound (plus a slight excess) should be dried over P_2O_5 in an open container for the length of time recommended in Appendix 3, column 4.

To prepare an oil solution containing one or several elements of interest, a solubilizing agent is prepared by placing 5 ml of 2-ethylhexanoic acid, 4 ml of 6-methyl-2,4-heptanedione, and 2 ml of xylene in a weighed 200-ml flask. The calculated weight of standard sample (Appendix 3, column 5) is carefully transferred to the flask. The suspension is gently heated on a hot plate, with swirling, until the salt has dissolved to a clear solution.

The salts are added, one at a time, and completely dissolved before the next compound is added. After each addition, the mixture is gently heated, tilted, and rotated in order to wash down any particles adhering to the inside of the flask. After the last compound has been added and dissolved, 2 ml of bis(2-ethylhexyl)amine is added. The beaker is now washed with six 15-ml portions of lubricating oil, and these rinsings are all transferred to the flask. This last step is omitted if the sample is transferred directly to the flask from the sample container, and not weighed into an intermediate beaker.

The oil solution is now allowed to cool to room temperature, and cool lubicating oil is added to the flask until the weight of the solution is 100 ± 0.5 g. The solution is reheated to 85°C, allowed to cool to room temperature, transferred to a storage bottle, and tightly stoppered.

Compounds of silver and mercury are sensitive to prolonged heating and so, they should be added last to any mixture. For the other compounds, the order of addition is immaterial.

Appendix 1: Standard Stock Solutions*

Element	Procedure
Aluminum	Dissolve 1.000 g Al wire in minimum amount of 2 M HCl; dilute to volume.
Antimony	Dissolve 1.000 g Sb in (1) 10 ml HNO_3 plus 5 ml HCl, and dilute to volume when dissolution is complete; or (2) 18 ml HBr plus 2 ml liquid Br_2, when dissolution is complete add 10 ml $HClO_4$, heat in a well-ventilated hood while swirling until white fumes appear and continue for several minutes to expel all HBr, then cool and dilute to volume.
Arsenic	Dissolve 1.3203 g of As_2O_3 in 3 ml 8 M HCl and dilute to volume; or treat the oxide with 2 g NaOH and 20 ml water, after dissolution dilute to 200 ml, neutralize with HCl (pH meter), and dilute to volume.
Barium	(1) Dissolve 1.7787 g $BaCl_2 \cdot 2H_2O$ (fresh crystals) in water and dilute to volume. (2) Dissolve 1.516 g $BaCl_2$ (dried at 250°C for 2 hr) in water and dilute to volume. (3) Treat 1.4367 g $BaCO_3$ with 300 ml water, slowly add 10 ml of HCl and, after the CO_2 is released by swirling, dilute to volume.

* 1000 μg/ml as the element in a final volume of 1 liter unless stated otherwise.

Appendix 1: (Cont.)

Element	Procedure
Beryllium	(1) Dissolve 19.655 g $BeSO_4 \cdot 4H_2O$ in water, add 5 ml HCl (or HNO_3), and dilute to volume. (2) Dissolve 1.000 g Be in 25 ml 2 M HCl, then dilute to volume.
Bismuth	Dissolve 1.000 g Bi in 8 ml of 10 M HNO_3, boil gently to expel brown fumes, and dilute to volume.
Boron	Dissolve 5.720 g fresh crystals of H_3BO_3 and dilute to volume.
Bromine	Dissolve 1.489 g KBr (or 1.288 g NaBr) in water and dilute to volume.
Cadmium	(1) Dissolve 1.000 g Cd in 10 ml of 2 M HCl; dilute to volume. (2) Dissolve 2.282 g $3CdSO_4 \cdot 8H_2O$ in water; dilute to volume.
Calcium	Place 2.4973 g $CaCO_3$ in volumetric flask with 300 ml water, carefully add 10 ml HCl, after CO_2 is released by swirling, dilute to volume.
Cerium	(1) Dissolve 4.515 g $(NH_4)_4Ce(SO_4)_4 \cdot 2H_2O$ in 500 ml water to which 30 ml H_2SO_4 had been added, cool, and dilute to volume. Advisable to standardize against As_2O_3. (2) Dissolve 3.913 g $(NH_4)_2Ce(NO_3)_6$ in 10 ml H_2SO_4, stir 2 min, cautiously introduce 15 ml water and again stir 2 min. Repeat addition of water and stirring until all the salt has dissolved, then dilute to volume.
Cesium	Dissolve 1.267 g CsCl and dilute to volume. Standardize: Pipette 25 ml of final solution to Pt dish, add 1 drop H_2SO_4, evaporate to dryness, and heat to constant weight at $\not> 800°C$. Cs (in μg/ml) = (40)(0.734)(wt of residue)
Chlorine	Dissolve 1.648 g NaCl and dilute to volume.
Chromium	(1) Dissolve 2.829 g $K_2Cr_2O_7$ in water and dilute to volume. (2) Dissolve 1.000 g Cr in 10 ml HCl, and dilute to volume.
Cobalt	Dissolve 1.000 g Co in 10 ml of 2 M HCl, and dilute to volume.
Copper	(1) Dissolve 3.929 g fresh crystals of $CuSO_4 \cdot 5H_2O$, and dilute to volume. (2) Dissolve 1.000 g Cu in 10 ml plus 5 ml water to which HNO_3 (or 30% H_2O_2) is added dropwise until dissolution is complete. Boil to expel oxides of nitrogen and chlorine, then dilute to volume.
Dysprosium	Dissolve 1.1477 g Dy_2O_3 in 50 ml of 2 M HCl; dilute to volume.
Erbium	Dissolve 1.1436 g Er_2O_3 in 50 ml of 2 M HCl; dilute to volume.
Europium	Dissolve 1.1579 g Eu_2O_3 in 50 ml of 2 M HCl; dilute to volume.
Fluorine	Dissolve 2.210 g NaF in water and dilute to volume.
Gadolinium	Dissolve 1.152 g Gd_2O_3 in 50 ml of 2 M HCl; dilute to volume.
Gallium	Dissolve 1.000 g Ga in 50 ml of 2 M HCl; dilute to volume.
Germanium	Dissolve 1.4408 g GeO_2 with 50 g oxalic acid in 100 ml of water; dilute to volume.

Appendix 1: (Cont.)

Element	Procedure
Gold	Dissolve 1.000 g Au in 10 ml of hot HNO_3 by dropwise addition of HCl, boil to expel oxides of nitrogen and chlorine, and dilute to volume. Store in amber container away from light.
Hafnium	Transfer 1.000 g Hf to Pt dish, add 10 ml of 9 M H_2SO_4, and then slowly add HF dropwise until dissolution is complete. Dilute to volume with 10% H_2SO_4.
Holmium	Dissolve 1.1455 g Ho_2O_3 in 50 ml of 2 M HCl; dilute to volume.
Indium	Dissolve 1.000 g In in 50 ml of 2 M HCl; dilute to volume.
Iodine	Dissolve 1.308 g KI in water and dilute to volume.
Iridium	(1) Dissolve 2.465 g Na_3IrCl_6 in water and dilute to volume. (2) Transfer 1.000 g Ir sponge to a glass tube, add 20 ml of HCl and 1 ml of $HClO_4$. Seal the tube and place in an oven at 300°C for 24 hr. Cool, break open the tube, transfer the solution to a volumetric flask, and dilute to volume. Consult (4) and observe all safety precautions in opening the glass tube.
Iron	Dissolve 1.000 g Fe wire in 20 ml of 5 M HCl; dilute to volume.
Lanthanum	Dissolve 1.1717 g La_2O_3 (dried at 110°C) in 50 ml of 5 M HCl, and dilute to volume.
Lead	(1) Dissolve 1.5985 g $Pb(NO_3)_2$ in water plus 10 ml HNO_3, and dilute to volume. (2) Dissolve 1.000 g Pb in 10 ml HNO_3, and dilute to volume.
Lithium	Dissolve a slurry of 5.3228 g Li_2CO_3 in 300 ml of water by addition of 15 ml HCl, after release of CO_2 by swirling, dilute to volume.
Lutetium	Dissolve 1.6079 g $LuCl_3$ in water and dilute to volume.
Magnesium	Dissolve 1.000 g Mg in 50 ml of 1 M HCl and dilute to volume.
Manganese	(1) Dissolve 1.000 g Mn in 10 ml HCl plus 1 ml HNO_3, and dilute to volume. (2) Dissolve 3.0764 g $MnSO_4 \cdot H_2O$ (dried at 105°C for 4 hr) in water and dilute to volume. (3) Dissolve 1.5824 g MnO_2 in 10 HCl in a good hood, evaporate to gentle dryness. dissolve residue in water and dilute to volume.
Mercury	Dissolve 1.000 g Hg in 10 ml of 5 M HNO_3 and dilute to volume.
Molybdenum	(1) Dissolve 2.0425 g $(NH_4)_2MoO_4$ in water and dilute to volume. (2) Dissolve 1.5003 g MoO_3 in 100 ml of 2 M ammonia, and dilute to volume.
Neodymium	Dissolve 1.7373 g $NdCl_3$ in 100 ml 1 M HCl and dilute to volume.
Nickel	Dissolve 1.000 g Ni in 10 ml hot HNO_3, cool, and dilute to volume.
Niobium	Transfer 1.000 g Nb (or 1.4305 g Nb_2O_5) to Pt dish, add 20 ml HF, and heat gently to complete dissolution. Cool, add 40 ml H_2SO_4, and evaporate to fumes of SO_3. Cool and dilute to volume with 8 M H_2SO_4.

Appendix 1: (Cont.)

Element	Procedure
Osmium	Dissolve 1.3360 g OsO_4 in water and dilute to 100 ml. Prepare only as needed as solution loses strength on standing unless Os is reduced by SO_2 and water is replaced by 100 ml 0.1 M HCl.
Palladium	Dissolve 1.000 g Pd in 10 ml of HNO_3 by dropwise addition of HCl to hot solution; dilute to volume.
Phosphorus	Dissolve 4.260 g $(NH_4)_2HPO_4$ in water and dilute to volume.
Platinum	Dissolve 1.000 g Pt in 40 ml of hot aqua regia, evaporate to incipient dryness, add 10 ml HCl and again evaporate to moist residue. Add 10 ml HCl and dilute to volume.
Potassium	Dissolve 1.9067 g KCl (or 2.8415 g KNO_3) in water and dilute to volume.
Praseodymium	Dissolve 1.1703 g Pr_2O_3 in 50 ml of 2 M HCl; dilute to volume.
Rhenium	Dissolve 1.000 g Re in 10 ml of 8 M HNO_3 in an ice bath until initial reaction subsides, then dilute to volume.
Rhodium	Dissolve 1.000 g Rh by the sealed-tube method described under iridium (4).
Rubidium	Dissolve 1.4148 g RbCl in water. Standardize as described under cesium. Rb (in $\mu g/ml$) = (40)(0.320)(wt of residue).
Ruthenium	Dissolve 1.317 g RuO_2 in 15 ml of HCl; dilute to volume.
Samarium	Dissolve 1.1596 g Sm_2O_3 in 50 ml of 2 M HCl; dilute to volume.
Scandium	Dissolve 1.5338 g Sc_2O_3 in 50 ml of 2 M HCl; dilute to volume.
Selenium	Dissolve 1.4050 g SeO_2 in water and dilute to volume or dissolve 1.000 g Se in 5 ml of HNO_3, then dilute to volume.
Silicon	Fuse 2.1393 g SiO_2 with 4.60 g Na_2CO_3, maintaining melt for 15 min in Pt crucible. Cool, dissolve in warm water, and dilute to volume. Solution contains also 2000 $\mu g/ml$ sodium.
Silver	(1) Dissolve 1.5748 g $AgNO_3$ in water and dilute to volume. (2) Dissolve 1.000 g Ag in 10 ml of HNO_3; dilute to volume. Store in amber glass container away from light.
Sodium	Dissolve 2.5421 g NaCl in water and dilute to volume.
Strontium	Dissolve a slurry of 1.6849 g $SrCO_3$ in 300 ml of water by careful addition of 10 ml of HCl, after release of CO_2 by swirling, dilute to volume.
Sulfur	Dissolve 4.122 g $(NH_4)_2SO_4$ in water and dilute to volume.
Tantalum	Transfer 1.000 g Ta (or 1.2210 g Ta_2O_5) to Pt dish, add 20 ml of HF, and heat gently to complete the dissolution. Cool, add 40 ml of H_2SO_4 and evaporate to heavy fumes of SO_3. Cool and dilute to volume with 50% H_2SO_4.
Tellurium	(1) Dissolve 1.2508 g TeO_2 in 10 ml of HCl; dilute to volume. (2) Dissolve 1.000 g Te in 10 ml of warm HCl with dropwise addition of HNO_3, then dilute to volume.

Appendix 1: (Cont.)

Element	Procedure
Terbium	Dissolve 1.6692 g of $TbCl_3$ in water, add 1 ml of HCl, and dilute to volume.
Thallium	Dissolve 1.3034 g $TINO_3$ in water and dilute to volume.
Thorium	Dissolve 2.3794 g $Th(NO_3)_4 \cdot 4H_2O$ in water, add 5 ml HNO_3, and dilute to volume.
Thulium	Dissolve 1.142 g Tm_2O_3 in 50 ml of 2 M HCl; dilute to volume.
Tin	Dissolve 1.000 g Sn in 15 ml of warm HCl; dilute to volume.
Titanium	Dissolve 1.000 g Ti in 10 ml of H_2SO_4 with dropwise addition of HNO_3; dilute to volume with 5% H_2SO_4.
Tungsten	Dissolve 1.7941 g of $Na_2WO_4 \cdot 2H_2O$ in water and dilute to volume.
Uranium	Dissolve 2.1095 g $UO_2(NO_3)_2 \cdot 6H_2O$ (or 1.7734 g uranyl acetate dihydrate) in water and dilute to volume.
Vanadium	Dissolve 2.2963 g NH_4VO_3 in 100 ml of water plus 10 ml of HNO_3; dilute to volume.
Ytterbium	Dissolve 1.6147 g $YbCl_3$ in water and dilute to volume.
Yttrium	Dissolve 1.2692 g Y_2O_3 in 50 ml of 2 M HCl and dilute to volume.
Zinc	Dissolve 1.000 g Zn in 10 ml of HCl; dilute to volume.
Zirconium	Dissolve 3.533 g $ZrOCl_2 \cdot 8H_2O$ in 50 ml of 2 M HCl, and dilute to volume. Solution should be standardized.

Appendix 2: Special Aqueous Stock Solutions

	Concentration	Procedure
Element		
Barium	0.0625 N	Dissolve 6.506 g $BaCl_2$ (dried at 250°C for 2 hr), dilute to volume. One ml is equivalent to 1.00 mg of sulfate sulfur.
Calcium	100 meq per liter	Dissolve a slurry of 5.006 g $CaCO_3$ (dried at 105°C) in 300 ml of water by careful addition of 10 ml HCl; after CO_2 is released, dilute to volume.
Lithium	200 meq per liter (as Li)	Dissolve a slurry of 7.392 g Li_2CO_3 in 300 ml of water by addition of 15 ml HCl; after release of CO_2, dilute to volume.
	2000 μg/ml (as Li_2O)	Dissolve 4.945 g Li_2CO_3 as above.

Appendix 2: (Cont.)

	Concentration	Procedure
Magnesium	100 meq per liter (as Mg)	Dissolve 1.2156 g Mg in 50 ml of 1 M HCl; dilute to volume.
Manganese	1000 μg/ml (as Mn_2O_3)	Dissolve 2.1412 g $MnSO_4 \cdot H_2O$ (dried at 105°C for 4 hr) and dilute to volume.
Potassium	100 meq per liter (as K)	Dissolve 7.456 g KCl and dilute to volume.
	1000 μg/ml (as K_2O)	Dissolve 1.5830 g KCl and dilute to volume.
Sodium	100 meq per liter (as Na)	Dissolve 5.846 g NaCl and dilute to volume.
	1000 μg/ml (as Na_2O)	Dissolve 1.886 g NaCl and dilute to volume.
Strontium	1000 μg/ml (as SrO)	Dissolve a slurry of 1.6848 g $SrCO_3$ as described for calcium.
Mixture		
Alkali-manganese solution	1000 μg/ml each Na_2O, K_2O, Mn_2O_3	Dissolve in water 1.886 g NaCl, 1.583 g KCl, and 2.1412 g $MnSO_4 \cdot H_2O$ (dried at 105°C for 4 hr) and dilute to volume.
Lime-acid solution	6300 μg/ml CaO in 1:9 HCl	Add 100 ml HCl to 11.244 g $CaCO_3$ suspended in 300 ml of water; after CO_2 is released, dilute to volume.
Sea water	Chlorinity 38%	Dissolve 47.0 g NaCl, 21.3 g $MgCl_2 \cdot 6H_2O$, 17.8 g $Na_2SO_4 \cdot 10H_2O$, 2.20 g $CaCl_2$, 1.46 g KCl, 0.38 g $NaHCO_3$, and 0.015 g $SrCl_2$; and dilute to volume.

Appendix 3: Standard Reference Materials in Lubricating Oil (Element of Interest, 500 µg/g)

NBS No.[a]	Element of interest	Compound	Drying time, hr[b]	Weight used, g[c]
1075a	Al	2-Ethylhexanoate	48	0.620
1051b	Ba	Cyclohexanebutyrate	24	0.174
1063a	B	Menthyl borate	2	2.08
1053	Cd	Cyclohexanebutyrate	48	0.202
1074a	Ca	2-Ethylhexanoate	48	0.373

Appendix 3: (Cont.)

NBS No.[a]	Element of interest	Compound	Drying time, hr[b]	Weight used, g^c
1078	Cr	Tris(1-phenyl-1,3-butanediono)-chromium(III)	1	0.515
1055b	Co	Cyclohexanebutyrate	24	0.337
1080	Cu	Bis(1-phenyl-1,3-butanediono)-copper(II)	0.5	0.303
1079b	Fe	Tris(1-phenyl-1,3-butanediono)-iron(III)	1	0.485[d]
1059b	Pb	Cyclohexanebutyrate	48	0.136
1060a	Li	Cyclohexanebutyrate	24	1.22
1061b	Mg	Cyclohexanebutyrate	48	0.765
1062a	Mn	Manganese(II) cyclohexanebutyrate	48	0.362
1064	Hg	Mercury(II) cyclohexanebutyrate	2	0.138
1065b	Ni	Cyclohexanebutyrate	48	0.360
1071a	P	Triphenyl phosphate	2	0.526
1076	K	Erucate	2	0.495
1066	Si	Octaphenylcyclotetrasiloxane	0	0.141[e]
1077a	Ag	2-Ethylhexanoate	24	0.117
1069b	Na	Cyclohexanebutyrate	48	0.417
1070a	Sr	Cyclohexanebutyrate	24	0.242
1057b	Sn	Dibutyltin bis(2-ethylhexanoate)	2	0.218
1052b	V	Bis(1-phenyl-1,3-butanediono)-oxovanadium(IV)	2	0.348
1073b	Zn	Cyclohexanebutyrate	48	0.299

[a] National Bureau of Standards (U.S.) standard reference material.
[b] Dried over P_2O_5 in desiccator.
[c] Weight of dried material which contains 50 mg of element of interest.
[d] Dried in oven at 110°C.
[e] Weight of element, 20 mg, and gives a final concentration of 200 $\mu g/g$.

REFERENCES

1. R. E. Michaelis (ed.), *Report on Available Standard Samples, Reference Samples, and High-Purity Materials for Spectrochemical Analysis, ASTM Data Series DS-2*, American Society for Testing Materials, Philadelphia, Pa., 1963.
2. *Catalog and Price List of Standard Materials Issued by The National Bureau of Standards, NBS Spec. Publ. 260*, U.S. Government Printing Office, Washington, D.C.

3. I. M. Kolthoff and P. J. Elving (eds.), *Treatise on Analytical Chemistry*, Part II (12 vols), Interscience-Wiley, New York.

4. W. F. Hillebrand, G. E. F. Lundell, H. A. Bright, and J. I. Hoffman, *Applied Inorganic Analysis*, 2nd ed., Wiley, New York, 1950, pp. 343–351.

5. H. S. Isbell, R. S. Tipson, J. L. Hague, B. F. Scribner, W. H. Smith, C. W. R. Wade, and A. Cohen, *Analytical Standards for Trace Elements in Petroleum Products*, National Bureau of Standards Monograph 54, U.S. Government Printing Office, Washington, D.C., 1962.

Author Index

Numbers in parentheses are reference numbers and indicate that an author's work is referred to although his name is not cited in the text. Numbers in italics show the page on which the complete reference is cited.

Subject Index

349

53